POGIL Activities for Introductory Anatomy and Physiology Courses

Murray Jensen, Project Lead
University of Minnesota

Anne Loyle, Project Assistant
University of Minnesota

Allison Mattheis, Activity Editor and Evaluation Lead
California State University Los Angeles

Contributors:

Sandy Craner
Southwest Minnesota State University

Joan Felice
Minneapolis Community and Technical College

Ron Gerrits
Milwaukee School of Engineering

Ann Henninger
Wartburg College

Jon Jackson
University of North Dakota

Michael Klug
Minneapolis Community and Technical College

Louise Millis
North Hennepin Community College

Ann Parsons
University of Wisconsin-Stout

Tinna Ross
North Hennepin Community College

Caleb Majerus
Contributing Artist

Funding for this project originated from the Minnesota State Colleges and Universities (MnSCU) system which provided funds and location for the initial workshop, and a grant from the National Science Foundation (DUE-1044221). The POGIL approach promotes a classroom environment in which students ask and answer questions, work in groups, and engage in discussions. For instructors who enjoy interacting with students, it's an ideal alternative to traditional lecture.

Acknowledgments

This project took over two years to complete and involved feedback, support and encouragement from many educators. We would like to thank everyone who participated in our workshops and the review process.

Jeff Adams
Roger Blevins
Wayne Busch
Tina Clarkson
Mary Ellen Doherty
Denise Ertl
Ann Marie Froehle
Mike Harris
Steve Johnson
Joe Kissner
Tracy Leider
Tangela Lou Marsh
Caroly Mitzuk
Melinda O'Connor
Shelly Paradies
Andrew Petto
Nicole Ruhland
Kelly Schuette
Nilanjana Senqupta Caballero
Roger Skugrud
Sarah Straud
Amanda Trewin
Lynette Youngsma
Michael Alfieri
Matt Brown
Laura Childs
Nancy Cripe
Joanne Dougherty
Linda Flora
Carol Gavareski
Julie Hood-Gonsalves
Mary Jurney
Jason LaPres
Ryan Lester
Tony McGee
Shari Morasko
Joshua Ohrtman
David Parkin
Alyson Purdy
Jay Sandal
Brianne Schumacher
Thomas Sharp
Tessa Spraetz
Sarah Suskovic
Jeanna Weldon
Brent Zabel
Carol Bischof
Patrick Brown
Julia Choate
Jessica Daniels
Rich Dhyanchand
Chad Flickbohm
Paul Gieser
Mary Januschka
John Kapsner
Charles Lawrence
Shari LitchGray
Kren McManus
Terence Morris
Michael Otto
Karen Perell-Gerson
David Rich
Roxanne Schmiesing
Susan Semmler
Mary Sinclair
Laura Stillwell
Lisa Tracy
Nita Worthley

Student Roles for POGIL Activities

Activity Title: ______________________________

Designated Group Size: ______________________________

In POGIL courses, most of the classwork is done in groups of three or four. The membership of the groups may change. The roles within a group will change to allow everyone an opportunity to try out a role. Here are some roles that are commonly used:

READER

Reads the activity out loud to the group. The reader must monitor their volume so that their group can hear them, but other groups are not disturbed. This helps keep everyone in the group together. The Manager/Facilitator will tell the Reader when it is time to read the next part of the activity.

RECORDER/SPOKESPERSON

Records the names and roles of the group members at the beginning of each activity. All students in the group should write down answers, but it is the recorder's job to decide what to record and is a log of the important concepts that the group has learned. This is frequently done after a group discussion. The recorder should also say things like, "Are we sure of this?" or "Does that sound accurate?" This person is responsible for reporting orally to the class when called for in class discussions.

MANAGER/FACILITATOR

Manages the group. Ensures that members are fulfilling their roles, that the assigned tasks are being accomplished on time and that all members of the group participate in activities and understand the concepts. This person keeps track of time, decides when the group should move on to the next question. They will also be responsible for looking things up in the textbook or the Internet when necessary.

REFLECTOR/QUALITY CONTROL

Observes and comments on group dynamics and behavior with respect to the learning process. The reflector/quality control may be called upon to report to the group (or entire class) about how well the group is operating (or what needs improvement) and why.

Note: Not all roles are assigned on any given day, and additional roles may be assigned to group members as needed.

Address inquiries to The POGIL Project, Box 3003, Lancster, PA 17604-3003.

ISBN: 9781118986745
SKY10022038_102820

Printed in the U.S.A.

Contents

Levels of Organization

NOTES FOR STUDENTS

- This activity will be completed in groups of three.
- Do not use the Internet or your textbooks unless instructed to do so.

PROCEDURE

1) Before beginning the activity, find three people you do not know and form a group of three.

2) Arrange yourselves in a circle and share the following introductory information:
   a) Name
   b) Favorite food
   c) Favorite subject or class
   d) Least favorite subject or class

3) Assign group roles:

- **READER**- this person will read out loud the text below.

- **RECORDER**- this person decides what to record as your group's answers. All students in the group should write down answers, but it is the recorder's job to decide what to record. This is frequently done after a group discussion. The recorder should also say things like, "Are we sure of this?" or "Does that sound accurate?" The recorder is also the group's spokesperson during class discussions.

- **FACILITATOR**-This person keeps track of time, decides when the group should move on to the next question, and promotes/encourages all group members to contribute to the group's discussion. They will also be responsible for looking things up in the textbook or on the Internet when necessary.

Model 1: Anatomy and Levels of Organization

An anatomist is a person who studies the structure of a living thing – how all the little things are organized into bigger things. The smallest living structures are cells, but there are things even smaller than a cell (such as atoms and molecules). Figure 1 shows the levels of organization used by anatomists. Table 1 names examples of each of the levels of organization shown in Figure 1

Figure 1: Levels of Organization in the Human Body

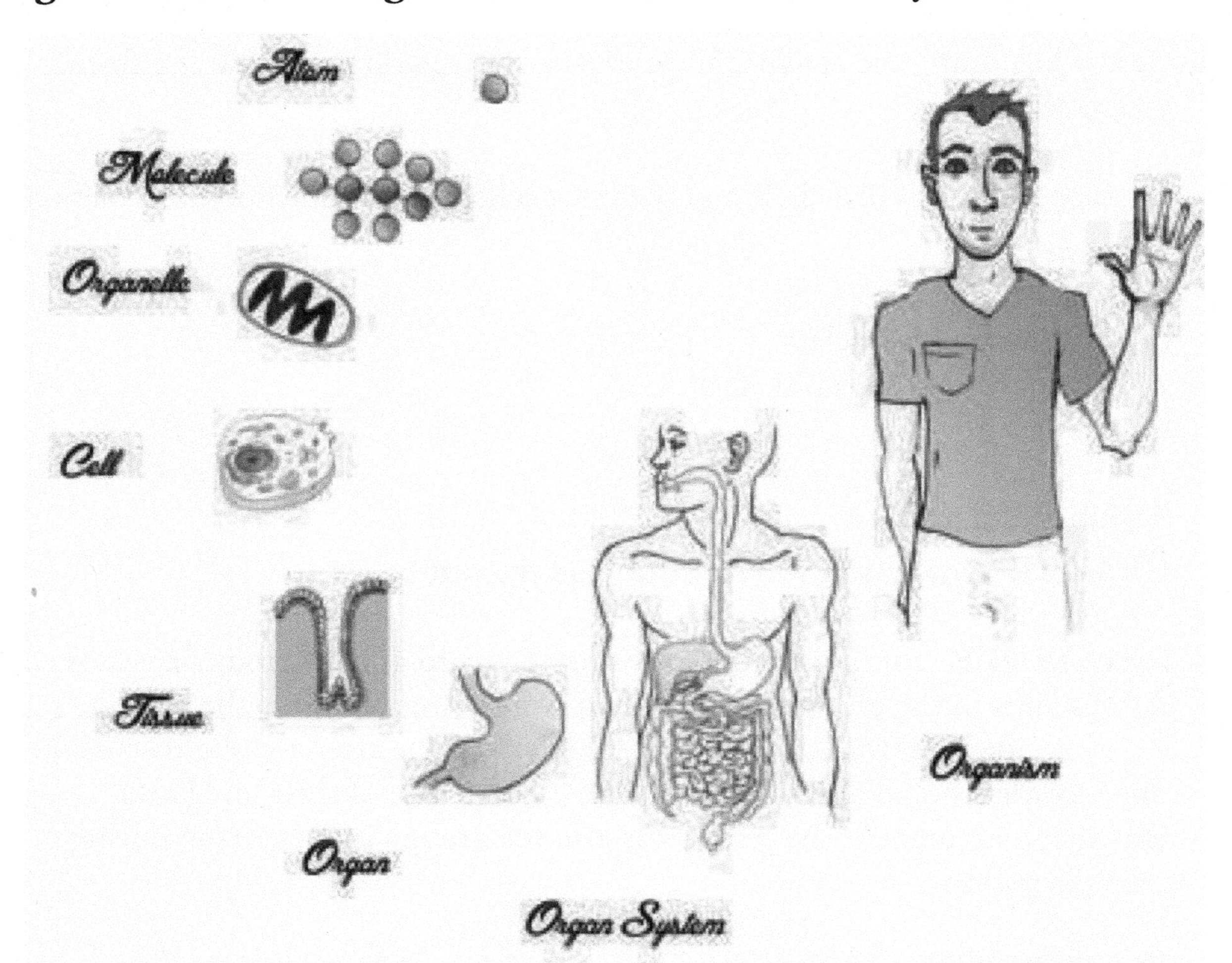

Level of Organization	Example
Atom	Carbon
Molecule	Carbon Dioxide
Cell	Stomach Cell
Tissue	Epithelial Tissue
Organ	Stomach
Organ System	Digestive System
Organism	YOU

QUESTIONS:

1. Using the list above, identify two terms that describe components that are smaller than cells and *four* that are larger than cells.

Smaller than a cell	Larger than a cell
1.	1.
2.	2.
	3.
	4.

2. What are two examples from the list above that can be found **inside** a cell?

3. Classify the follow images and label with the appropriate level of organization.

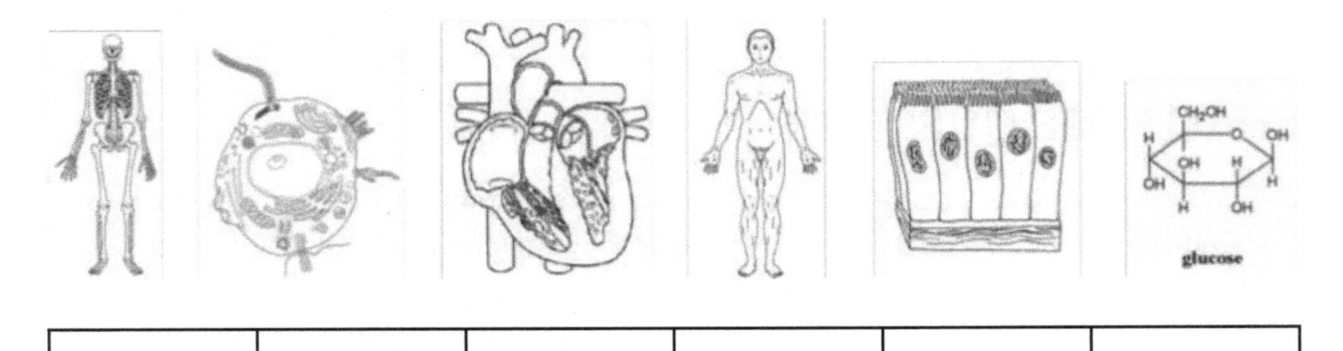

4. Circle "TRUE" or "FALSE" for the following statements:

a) Organs are composed of multiple tissues types. TRUE or FALSE
b) Tissues are composed of multiple cell types. TRUE or FALSE
c) Tissues are composed of multiple organ types. TRUE or FALSE

5. Spend 60 seconds working individually to write definitions for the two terms below. After 60 seconds, discuss your definitions with the group and decide who has the best definition.

a) Tissue:

b) Organ:

6. Without using books or the Internet, complete the chart by identifying the organ systems associated with the example organs listed: *(the first row is completed as an example. If you cannot identify the organ system, leave the box blank)*

Organ	Organ System
Femur	Skeletal System
Heart	
Brain	
Skin	
Pituitary Gland	
Lungs	
Stomach	
Spleen and Appendix	
Uterus	
Kidney	
Biceps Brachii	

CHALLENGE QUESTIONS:

7. Organs are sometimes shared by two or more systems - for example, your mouth can be considered a part of both the digestive and the respiratory systems. *Without* using your book or the Internet, try to name 3 organs that are shared by two or more body systems, and identify those body systems.

Example: Organ: *Mouth* What two systems? *Digestive & Respiratory Systems*
1: Organ: What two systems?
2: Organ: What two systems?
3: Organ: What two systems?

8. With your group, consider the following statements and determine if they are true or not. If they are NOT true, describe why (list exceptions that exist).

a) Within the body, all atoms combine to form molecules.

b) Within the body, all molecules in the body can be found inside cells.

Introduction to Medical Terminology

Model 1: The Anatomy of a Medical Term

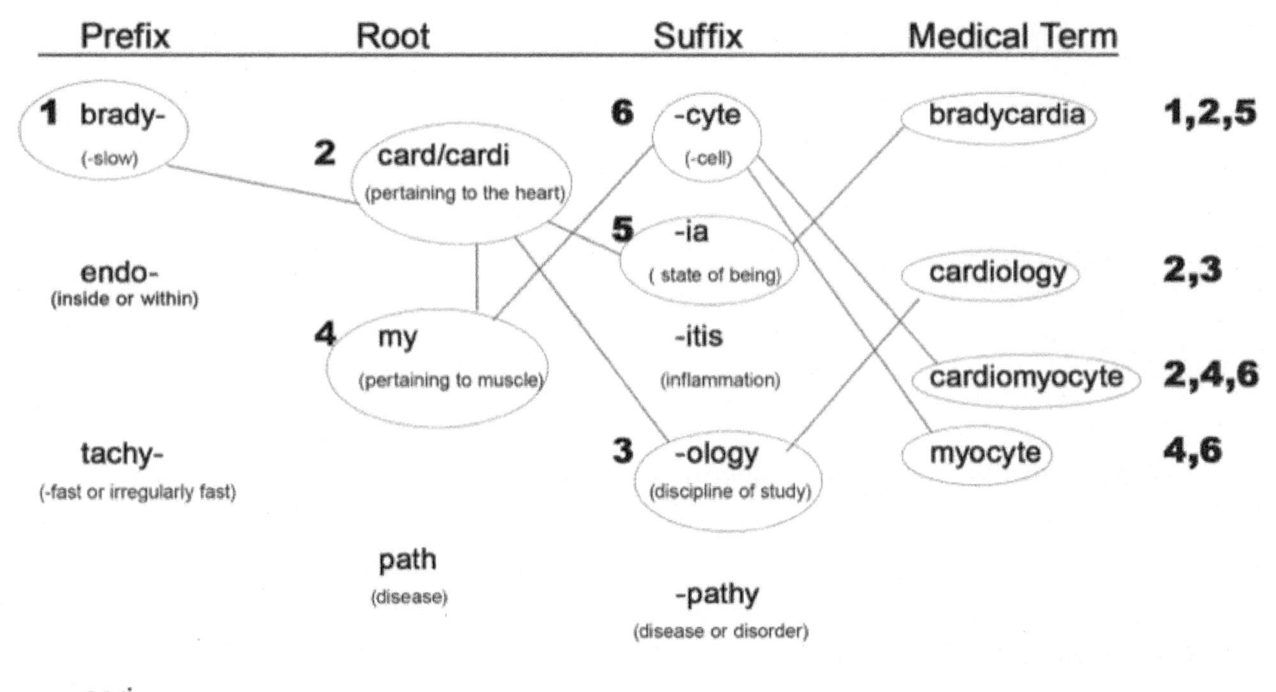

Figure 1: Prefix, Root, Suffix, Medical Term

QUESTIONS:

1. Use Figure 1 to determine the meaning of the following terms:

   a) cardiology:

   b) bradycardia:

   c) myocyte:

2. Use Figure 1 to build a word for the following definitions:

   a) The study of diseases-

   b) An irregular, fast heart rate-

3. Along with ***prefixes***, ***roots*** and ***suffixes***, medical terms can also include ***connecting vowels***. Referring to Model 1, circle the connecting vowels in the following terms:

   a) myocyte

   b) cardiology

   c) cardiomyocyte

4. Is it possible to have a medical term with multiple roots? If so, give an example.

5. Working individually for 60 seconds, try to list three medical professions whose titles contain a medical root term. (The word "cardiologist," for example, contains a root term, but the word "nurse" does not). After 60 seconds, compare your list with the other members of the group. (Key: these answers cannot be generated only by looking at Model 1.)

   a) cardiology – "cardio" means heart.

   b)

   c)

   d)

6. Modern medical terminology is based on two "dead" languages: ancient Greek and Latin. As a group, discuss the following two questions and write a one or two sentence explanation.

   a) What is a "dead" language?

   b) Why is "modern" medical terminology based on "dead" languages?

EXTENSION QUESTIONS:

If an item or event is named after a person it is called an **eponym**. For example, *Martin Luther King Jr. Day* is named after Martin Luther King, Jr. Another example is *saxophone*, which is named after the Belgian instrument designer and musician, Adolf Sax (see image). In medical terminology, eponyms are slowly being replaced by more descriptive medical terms. For example, the *islets of Langerhans*, a small island of cells in the pancreas are named after the German pathologist Paul Langerhans, is being replaced with the more descriptive term *pancreatic islets*. Medical terminology changes slowly, however, and many medical professionals still refer to the pancreatic islets as the islets of Langerhans.

7. List two eponyms that are NOT associated with medicine.

   a) Teddy bear: named after US President Teddy Roosevelt, who was an avid outdoorsman

   b)

   c)

8. As a group, try to list three medical eponyms and, if possible, also identify their locations and/or functions. *(One example, but not the location or function, has been given)*

   a) Achilles's tendon:

   b)

   c)

9. Without using the Internet, try to determine the meaning of the following terms *(Hint: start by circling and defining, all the root terms):*

   a) Electroencephalography

   b) Pneumonoultramicroscopicsilicovolcanoconiosis

10. Is it possible to have a medical term without a root? If so, try to think of an example. *(Think for 30 seconds, then discuss with your group)*

Introduction to Homeostasis

Model 1: Thermostat-controlled Heating System

This model shows a heating system for maintaining home/room temperature in cold weather. Most people consider a value around 23 degrees Celsius to be comfortable.

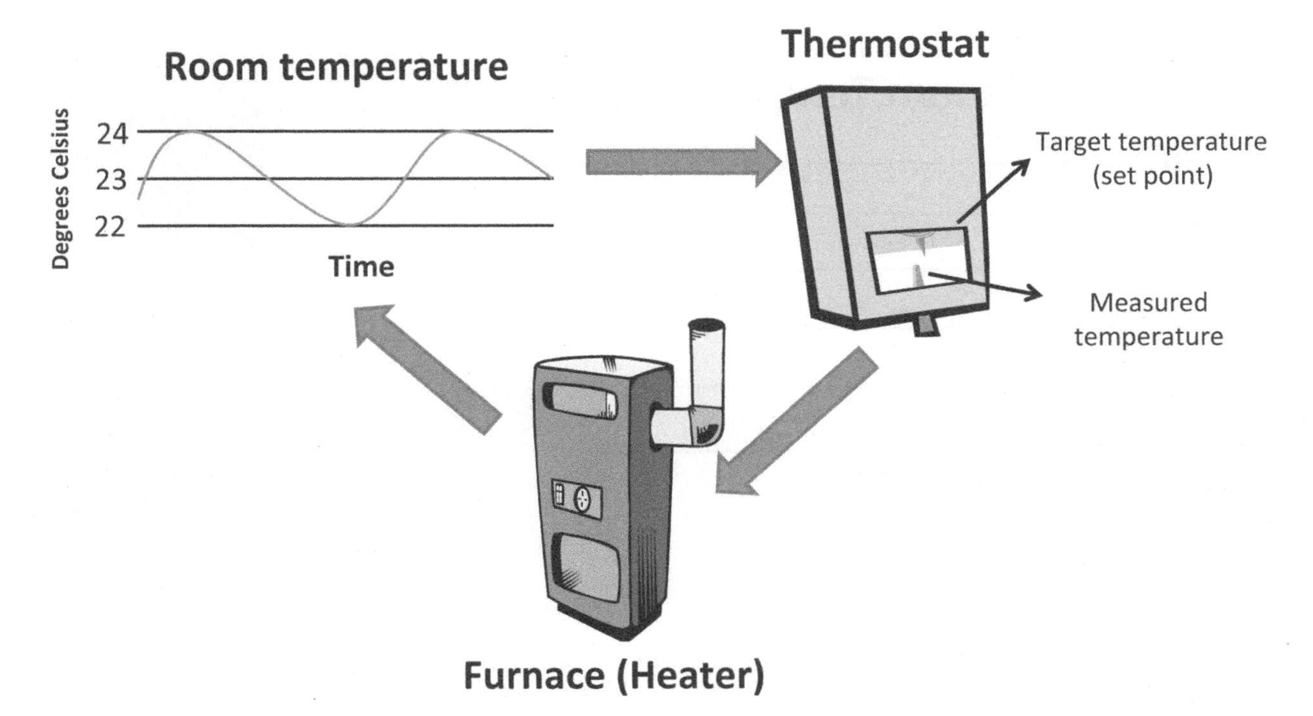

QUESTIONS:

1. What is the temperature range within the house over the time period shown?

2. What two values are used by the thermostat in its functioning?

3. At what temperature does the furnace turn on? At what temperature does it turn off?

4. In this scenario, what is the most likely value of the target temperature? Explain your reasoning.

5. Instead of using a furnace to heat a room, a propane heater could be used. If the room starts at 20 degrees Celsius and someone turns on the propane heater and leaves, plot the temperature of the room over time.

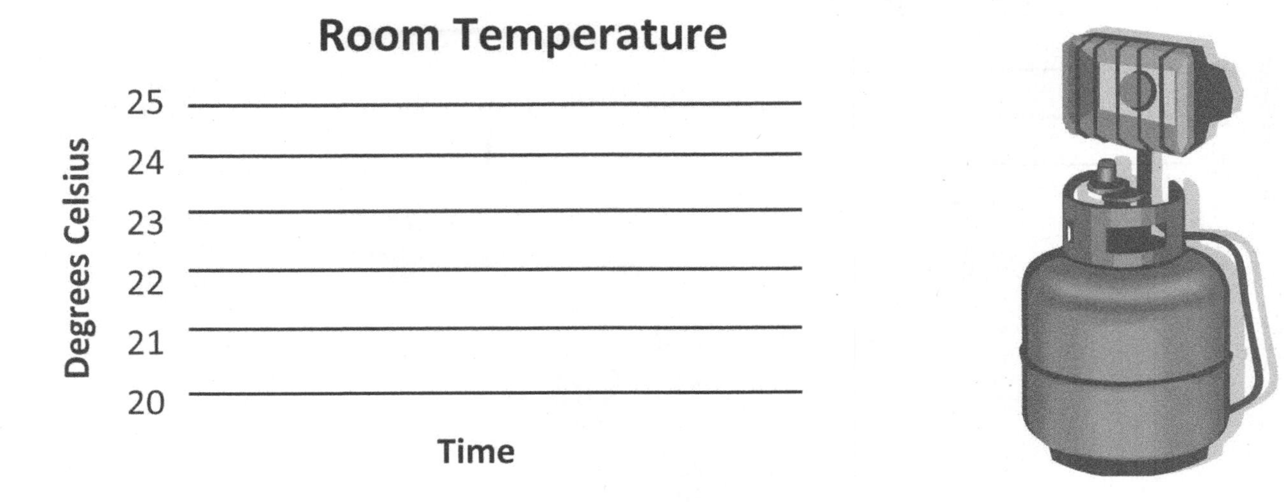

6. In contrast to the propane heater in question 5, the heating system shown in Model 1 uses "feedback" to maintain a target temperature. Individually, write a definition of *feedback* and then compare it with your team members' definitions.

7. Explain the benefit of a system like that shown in Model 1 that uses feedback to regulate temperature.

8. Using a grammatically correct sentence, write your best group definition for *feedback system*.

STOP: Discuss the definition of a feedback system with your instructor and/or class before moving on.

9. The specific type of feedback used in model 1 is *negative feedback*. This determination is based on how the system, or "loop," responds to a detected difference between the measured temperature and the target temperature. With regard to this difference, what is the goal of a negative feedback system?

10. The homeowner is growing Northern Australian orchids in the house. These orchids only grow in temperatures above 22 degrees Celsius. If the thermostat or the wire from the thermostat to the furnace breaks, what will happen to the temperature in the house? What will happen to the orchids? (Assume that it is winter in a northern climate and it is cold outside).

11. In the summer, when the weather gets hot, how could this loop be modified (settings or components) to keep the house around the same target temperature?

12. What is the most likely cause of a change in the target temperature (set point)? Is this change determined by one of the components of the loop itself?

Model 2: Homeostatic Feedback Loop

Negative feedback loops are used in living organisms and are similar to the system that controls the temperature in a house with a thermostat. Model 2A shows a generic model of a feedback loop and Figure 2B shows a sample loop important in maintaining blood pressure in the body. Models 2A and 2B also indicate variable levels can be affected by stimuli from outside of the loop, and that these variables impact function.

Model 2A: General Homeostatic Feedback Loop

Feedback Loop
Stimulus
Sensor (Receptor)
Information flow
Other information
Variable
Integrator (Comparator)
Effector(s)
Information flow
Function

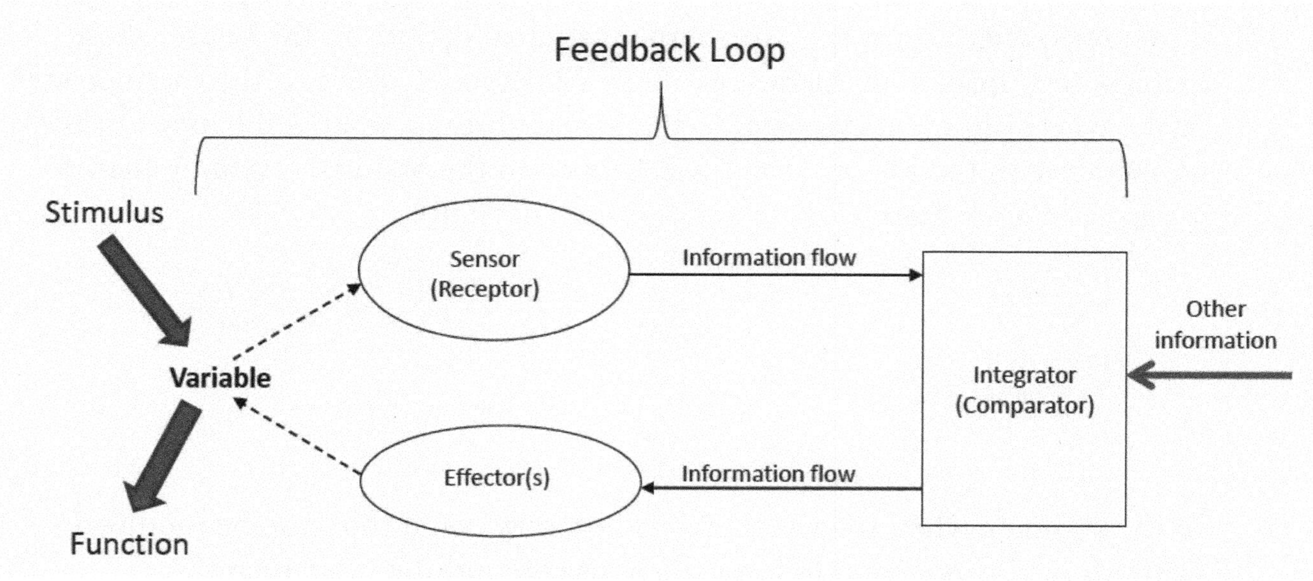

Model 2B: A Homeostatic Feedback Loop for the Regulation of Blood Pressure

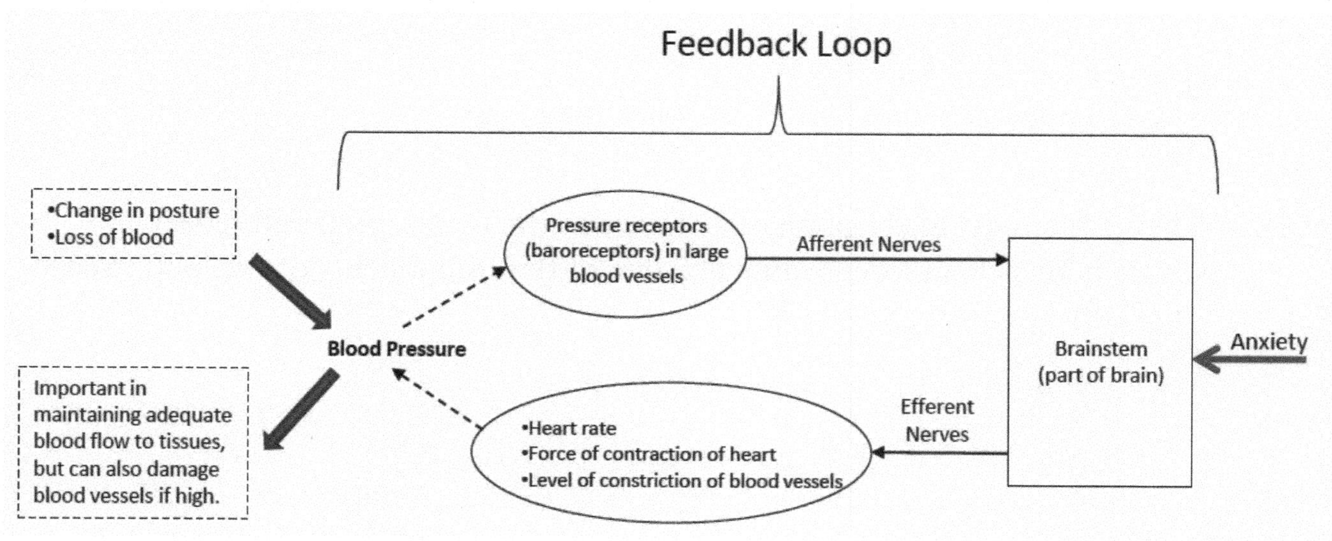

QUESTIONS (Refer to Models 1 and 2 for these questions):

13. Negative feedback loops contain three *functional components* that are necessary for regulating a variable (note that in this case the variable is not considered a functional component of the loop). Complete the table below by listing the three functional components of a feedback loop (add them in column 1 under Variable) and using information from Models 1 and 2 to fill in each box.

Variable and components of a feedback loop	Items from Model 1 that represent this aspect / component	Items from Model 2B that represent this aspect / component	Function of each aspect/ component
Variable	Room temperature		

14. In terms of biological function or process, what do the solid arrows shown in Model 2A represent?

15. In Model 2B, what is one stimulus for a change in blood pressure?

16. What is an example of *other information* that might determine the *set point* that the brainstem uses?

17. Based on model 2, individually write your best definition of *stimulus*.

18. If blood pressure is not maintained within a normal range, what consequences might result? (Complete the sentences below:)

    a) If blood pressure is too low,

    b) If blood pressure is too high,

19. Assuming that the feedback loop in Model 2B is functioning properly, predict what will happen to heart rate (HR) if the following changes occur:

    a) Blood pressure decreases:

    b) Blood pressure increases:

20. The goal of homeostasis is to prevent conditions such as those listed in the previous question from occurring. Individually, write your best definition of *homeostasis* and then compare answers with your team.

21. As a team, write your best definition of homeostasis.

STOP: Discuss the definition of a feedback system with your instructor and/or class before moving on.

22. Does a homeostatic feedback loop only function when there is a change from the set point? Explain.

23. Is the goal of a homeostatic feedback loop to maintain a regulated variable at a constant value? Explain your answer, referring to your definition of homeostasis.

24. Is a negative feedback loop limited to one effector? Explain (Hint: it may be helpful to refer to Figure 2b).

25. Does the integrator have to be a separate anatomical structure from the sensor? Explain (Hint: it may be helpful to refer to Model 1).

26. In Model 2, is heart rate (HR) a homeostatically controlled variable? Explain.

27. In addition to blood pressure, what other variables in the human body are controlled by homeostatic feedback loops? List two examples:

28. What happens when a negative feedback loop is damaged or interrupted? If possible, provide a specific example based on the variables listed in question 18.

29. The most common feedback loops in the body use the principle of negative feedback. But in some cases, positive feedback occurs for a limited amount of time. One example of this is in the formation of certain clotting factors when blood clotting is initiated. In this scenario the body needs a large number of clotting factors created in a short amount of time. How would a positive feedback loop differ from a negative feedback loop in terms of functional components and responses?

Membrane Transport

Model 1: Active Versus Passive Transport

Body fluids are solutions of water and dissolved solutes (ions, glucose, amino acids, etc.) that surround our cells. The plasma membrane is a selectively permeable barrier that allows some ions and molecules to pass through but prevents the passage of many molecules. The passage of some of these molecules through the semipermeable plasma membrane is essential for life. **Active transport** and **passive transport** are two mechanisms responsible for moving these necessary substances across the plasma membrane.

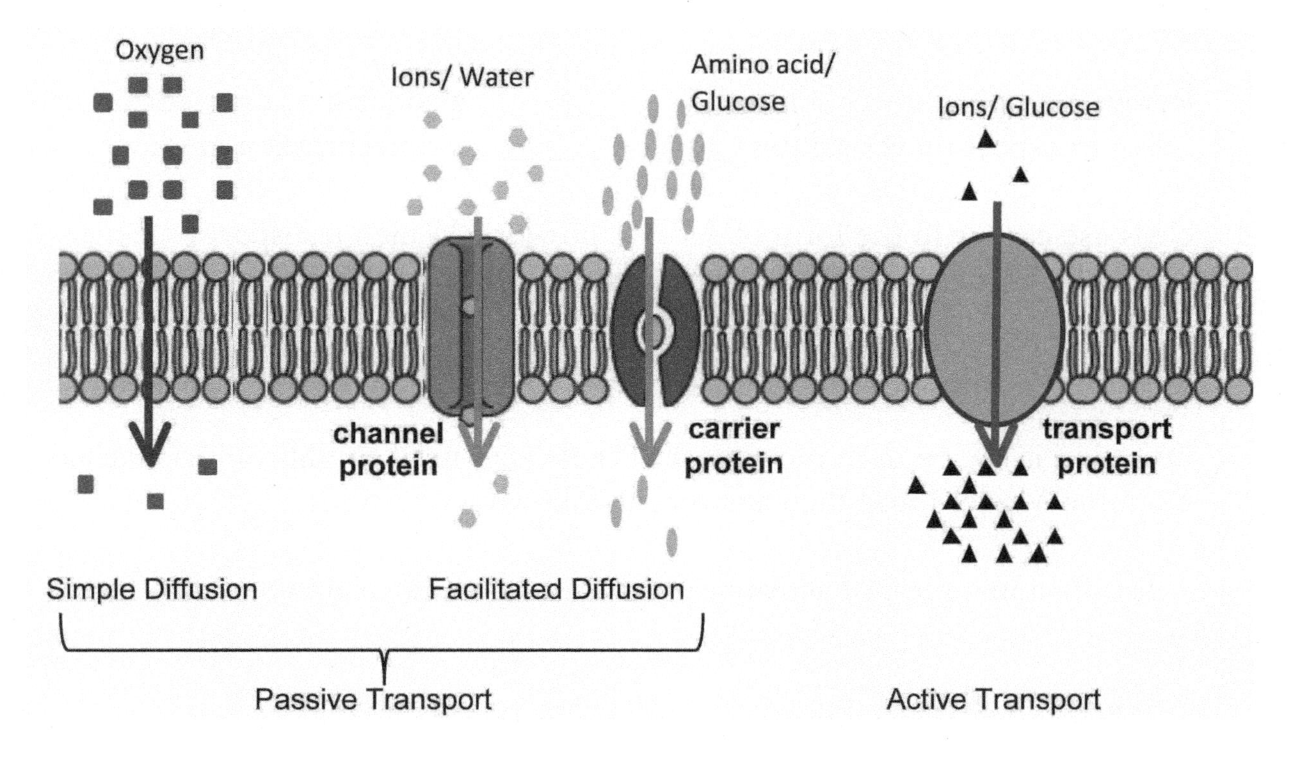

QUESTIONS:

1. Examine the Model above and identify the different transport processes. For each transport process, indicate the different amounts of solutes on each side of the membrane by labeling **"high solute concentration"** and **"low solute concentrate"** on the appropriate sides.

2. Transport mechanisms are defined by how solutes move relative to a **concentration gradient**, (i.e. high to low, low to high). Examine Model 1 and answer the following:

   a) In passive transport, how do solutes move along their concentration gradient?

   b) In active transport, how do solutes move along their concentration gradient?

3. Molecules are often described as moving **UP** or **DOWN** the concentration gradient. Fill-in the blanks:

   Passive transport moves solutes ______________ a concentration gradient, and active transport moves solutes ______________ a concentration gradient.

4. Cells use energy in the form of ATP to do work. Which transport process(es) use ATP? As a group, write a complete sentence to explain why.

5. Diffusion is the random movement of molecules until equilibrium is reached. Examine Model 1 and then answer the following questions:

   a) Is diffusion a form of active or passive transport? Explain your answer.

   b) What two types of diffusion are shown in the Model?

   c) Which process(es) involve transport directly through the phospholipid membrane?

   d) Which process(es) require protein channel/carrier for transport?

   e) Are solutes transported by <u>simple diffusion</u> **permeable** or **impermeable** to the plasma membrane?

   f) Are solutes transported by <u>facilitated diffusion</u> **permeable** or **impermeable** to the plasma membrane?

6. The rate of diffusion through a membrane is influenced by several factors. The Manager should have each member of the team take turns predicting which condition will allow diffusion to occur *faster* (circle one answer for each):

   - Size of molecules - **smaller** or **larger**?
   - Temperature of molecules - **cooler** or **warmer**?
   - Membrane surface area – **less** or **more**?
   - Membrane permeability – **less** or **more**?
   - Steepness of concentration gradient – **less different** or **more different**?

7. The Na-K pump actively transports potassium into the cell and sodium out of the cell across the plasma membrane. Draw a cell showing the Na and K distribution (high/low) for both the inside and outside of the cell.

8. A cell uses valuable energy to run active transport. Why does a cell need active transport? As a group, write an explanation that includes two examples of when active transport is used in the body.

Model 2: Osmosis

Osmosis is the diffusion of water through a semipermeable membrane to maintain equilibrium of solutes on both sides of the membrane. A solute imbalance causes water to move across the plasma membrane. The saying goes, "Water follows salt (solute)."

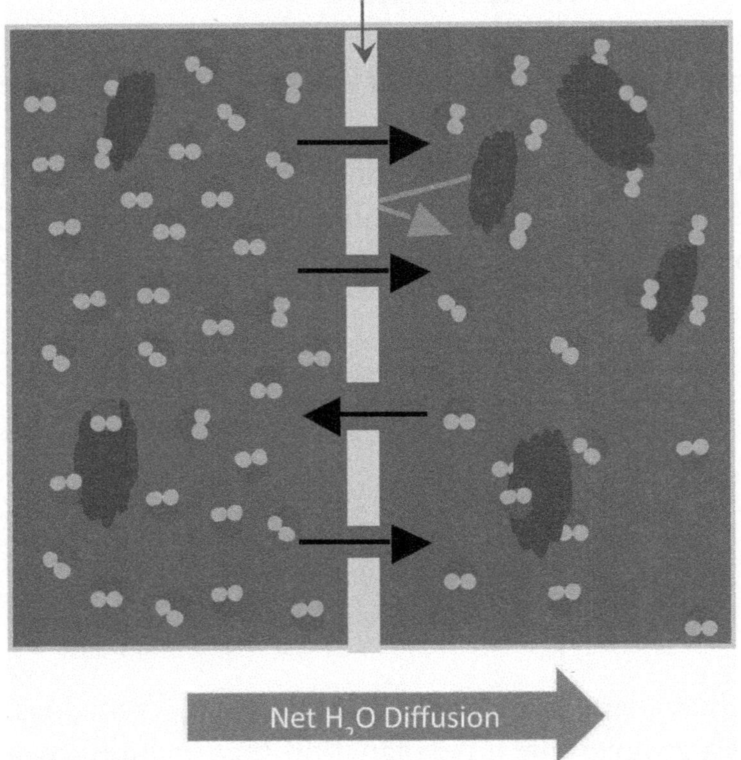

QUESTIONS:

9. On the diagram above, label the **water** and **solute** molecules.

10. Based on the model:

    a) How does water move relative to its <u>own concentration</u> gradient?

    b) How does water move relative to the <u>solute concentration</u> gradient?

11. What causes an imbalance in water concentration or solute concentration that leads to osmosis?

12. Water, a polar molecule, needs to constantly move in and out of the cell at rapid speed to maintain solute equilibrium. As a group, decide the specific type of transport used for osmosis and write an explanation using the terminology learned in Model 1:

13. The diagram below shows two U-tubes, which have two "arms" separated by a membrane that is impermeable to glucose at the bend of the tube. In Tube A, the left arm contains 6 glucose molecules and the right arm contains 14 glucose molecules. Assuming no additional solution is added to Tube B, show the volume and number of glucose molecules that would be present in each arm once equilibrium is reached:

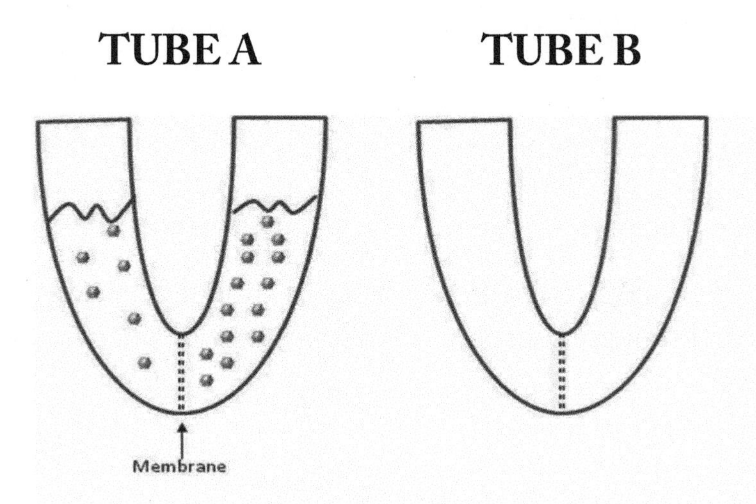

14. At equilibrium, how does the glucose concentration in the left side compare to the right side? *(Remember that concentration is the amount of solute per water)*

15. Imagine another scenario in which the two arms of Tube A are separated by a membrane that is permeable to sucrose. The left arm contains 2 sucrose molecules and the right arm contains 12 sucrose molecules. Draw a picture that predicts the volume of the solution and the concentration of sucrose molecules in each arm once equilibrium is reached.

Model 3: Tonicity

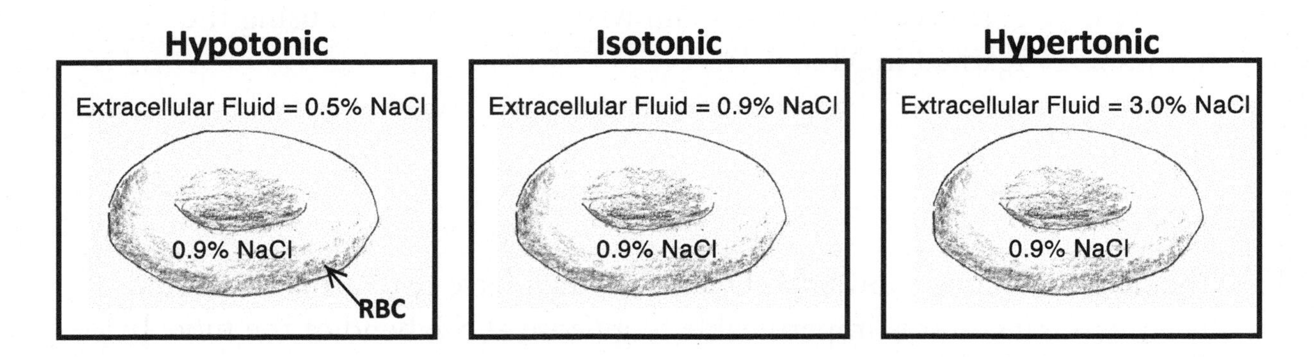

QUESTIONS:

16. Based on the information shown in Model 3, how does the impermeable solute concentration in the fluid surrounding the cell (the solution) compare to the inside of the cell? *(Use the words **less than, the same,** or **more than**)*

    a) Hypotonic:
    b) Isotonic:
    c) Hypertonic:

17. Based on your knowledge of osmosis (refer to model 2), predict the net water movement for each tonicity solution. *(Use the words **into, out of** or **no movement**)*

    a) Hypotonic:
    b) Isotonic:
    c) Hypertonic:

18. **Tonicity** is the ability of a solution to change the shape (tone) of a cell by altering its internal water volume. Based on your answer above, predict the shape of each RBC after it has reached equilibrium with each solution. Manager: have team members take turns predicting the final shape and agree as a group on a final answer for each solution.

    Hypotonic Isotonic Hypertonic

19. Is tonicity due to a solution containing a permeable or impermeable solute? Discuss as a group and justify your answer using information from Model 3 and your answers to questions 13 and 15.

20. Normal plasma osmolarity is about 300 mosm/kg. An elderly woman with diabetes has elevated blood glucose levels that raise her blood osmolarity to 310 mosm/kg In this diabetic patient, predict the water movement, if any, of the extracellular fluids and the shape of her red blood cells. *(Hint: the diabetic cells are initially in an isotonic fluid that is the same as normal blood)*

21. A patient was in a serious accident that caused a lot of blood loss. In an attempt to replenish body fluids, a volume of distilled water with a pH of 7.0 equal to that of the blood lost is transferred directly into the patient's veins. What will be the most probable impact on the person's red blood cells from this transfusion? Explain your answer using membrane transport terminology learned in these models:

22. Fill in the summary table below:

Transport Mechanism	Gradient Direction	Protein Carrier?	ATP needed?	Molecule Examples
Simple Diffusion				
Facilitated Diffusion				
Osmosis				
Active Transport				

Epithelial Tissue Histology

Model 1: Structure of Epithelial Tissue

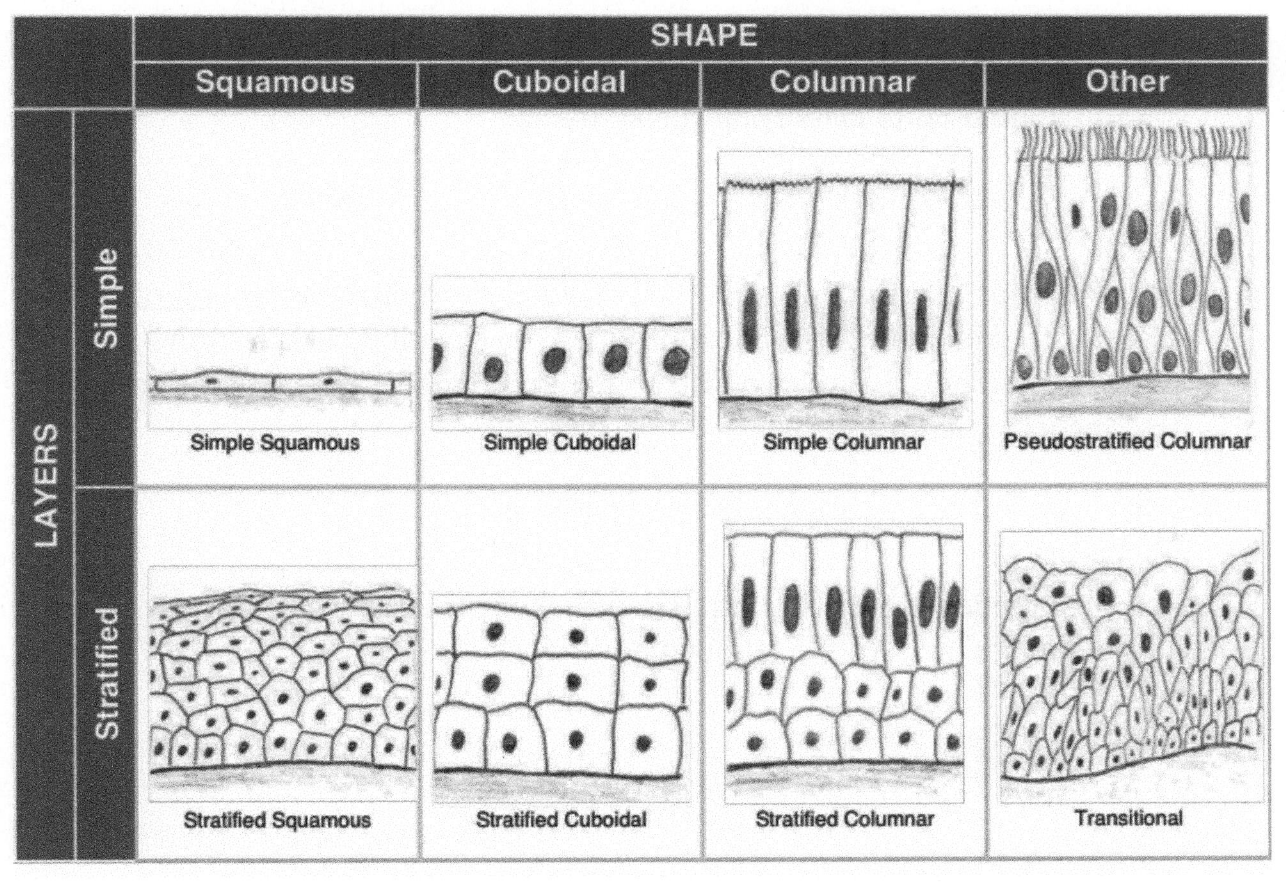

SHAPE
Squamous
Cuboidal
Columnar
Other
LAYERS
Simple
Stratified
Simple Squamous
Simple Cuboidal
Simple Columnar
Pseudostratified Columnar
Stratified Squamous
Stratified Cuboidal
Stratified Columnar
Transitional

QUESTIONS:

1. Describe the morphology (shape) of the following epithelial tissue cells.

   a) **Cuboidal**:

   b) **Columnar**:

   c) **Squamous**:

2. Write a grammatically correct sentence to describe the difference between **simple** and **stratified** epithelial tissue:

3. Reviewing your answers to questions #1 and #2, what two characteristics do you think are used to classify epithelial tissue?

4. Based on your examination of Model 1, which epithelial tissues do not conform to the naming classification rules you described in question #3? (*The manager* should have two different team members state a tissue and describe its structure).

5. Compare pseudostratified columnar epithelium and simple columnar epithelium and answer the following questions:

   a) How are these two tissues the same?

   b) What differentiates pseudostratified columnar epithelium from simple columnar?

6. The bottom (**basal**) layer of stratified epithelial tissue may have different cell shapes from the top (**apical**) layer. Which layer determines the shape classification of the tissue? Provide evidence to support your answer.

7. Name the specific type of epithelium described by the following: a tissue that consists of multiple layers of cells, in which the top layer is composed of columnar cells.

8. Before answering as a group, each person should individually complete this question by themselves. Once everyone is done, complete Question #9 as a group.

   a. Label the **apical** surface on each epithelial tissue photomicrograph.
   b. Label the **basal** surface on each epithelial tissue photomicrograph.
   c. Draw a bracket to indicate the location of the epithelial tissue.
   d. Name the specific epithelial type under both tissue photomicrographs.

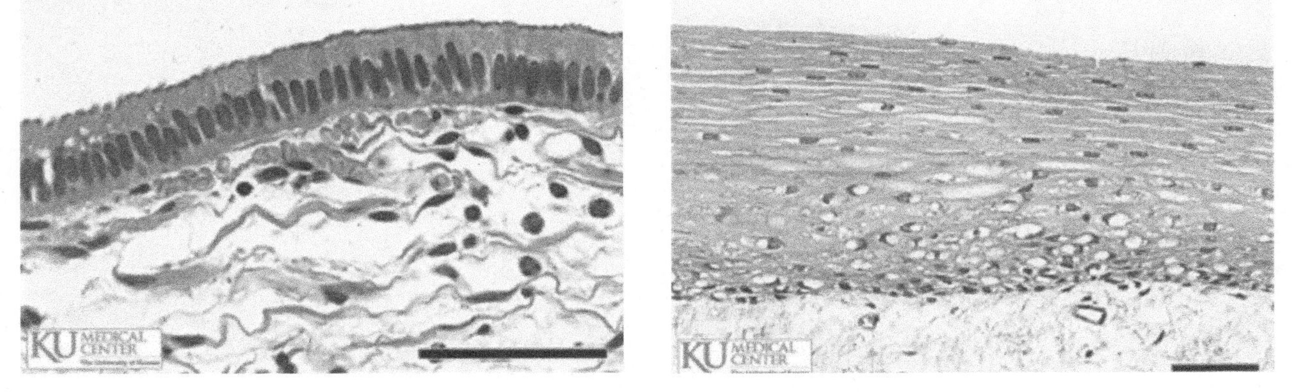

Images courtesy of The University of Kansas

Tissue = ______________________ Tissue = ______________________

9. Discuss and compare your individual answers to the above question with your group. (The manager should ensure that all group members share their answers). Develop a group consensus and write the best answer below:

10. The majority of dust is composed of human skin cells. What does this indicate about the rate of mitosis for epithelial tissue?

11. Epithelial cancers are the most common types of cancer. Answer the following questions together as a group:

    a) Briefly explain what cancer is.

    b) What do you think makes skin so susceptible to uncontrolled cellular growth?

PTH, Osteoporosis, and Calcium Homeostasis

Scenario

Your cousin, Suzanne, an outside hitter for the University of Kentucky volleyball, fell during practice. She was transported to the emergency room because she was in extreme pain and couldn't support her weight. Upon examination, the doctor explains that Suzanne has suffered a hip fracture. Because Suzanne is young, it is important to determine the underlying cause of the fracture. A number of tests have been ordered.

Model 1: Test Results

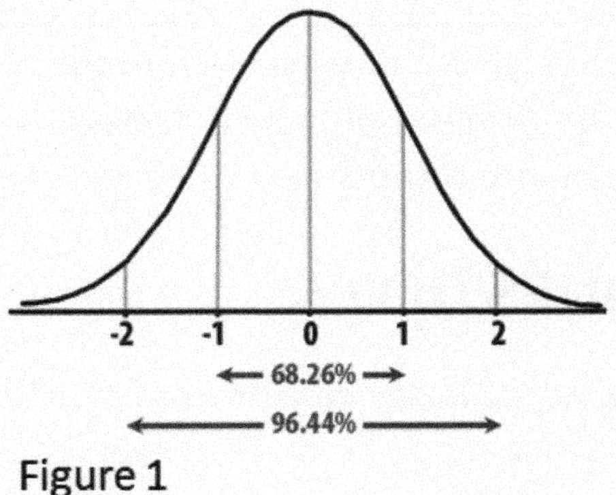

Figure 1

Bone mineral density (BMD) is measured using a special X-ray. A person's individual results are analyzed to yield what is called a T-score, the number of standard deviations (SD) above or below the mean (Figure 1), for a healthy adult of the same age, sex, and ethnicity as the patient. Sixty-eight percent of the population falls within one SD of the mean and 96% of the population will be within two SD of the mean. Table 1 explains what BMD T-scores mean.

Table 1. Interpretation of T-scores for BMD

T-score	What the T-score indicates
Greater than -1	Normal bone density
Between -1 and -2.5	Early signs of osteoporosis (osteopenia)
Less than -2.5	Osteoporosis

Table 2. Blood Test Results

	Normal Serum Values for Females	Suzanne's Serum Values
Parathyroid hormone	10-65 pg/ml [Conventional units] 10-65 ng/L [SI units]	128 pg/ml
Calcium	9.0-10.5 mg/dL [Conventional units] 2.2-2.6 mmol/L [SI units]	11.7 mg/dL
Calcitonin	< 6.4 pg/ml [Conventional units] < 6.4ng/L [SI units]	2.8 pg/ml

*serum values were taken from the Normal Laboratory Values MKSAP®14 (http://www.acpinternist.org/weekly/archives/2008/7/1/mksap.pdf) and Clinician's Ultimate Reference (http://www.globalrph.com/index.htm)

QUESTIONS:

Answer all questions with complete sentences

1. With your group members, discuss T-scores. Can negative T-scores be considered normal? Why or why not? (*Hint: use information from both the table and the text to answer this question)*

2. Suzanne receives a T-score of -3.6 in her bone mineral density (BMD) test results. What does this score indicate to her doctor?

3. Which items in Suzanne's blood test are:
   a) Within the normal range? ______________________________
   b) Above the normal range? ______________________________
   c) Below the normal range? ______________________________

4. Without using outside resources, what does your group suppose is the relationship between parathyroid levels and calcium levels in the blood?

Model 2: Histology of Bone Tissue

The doctor diagnoses Suzanne with osteoporosis.

Figure 2

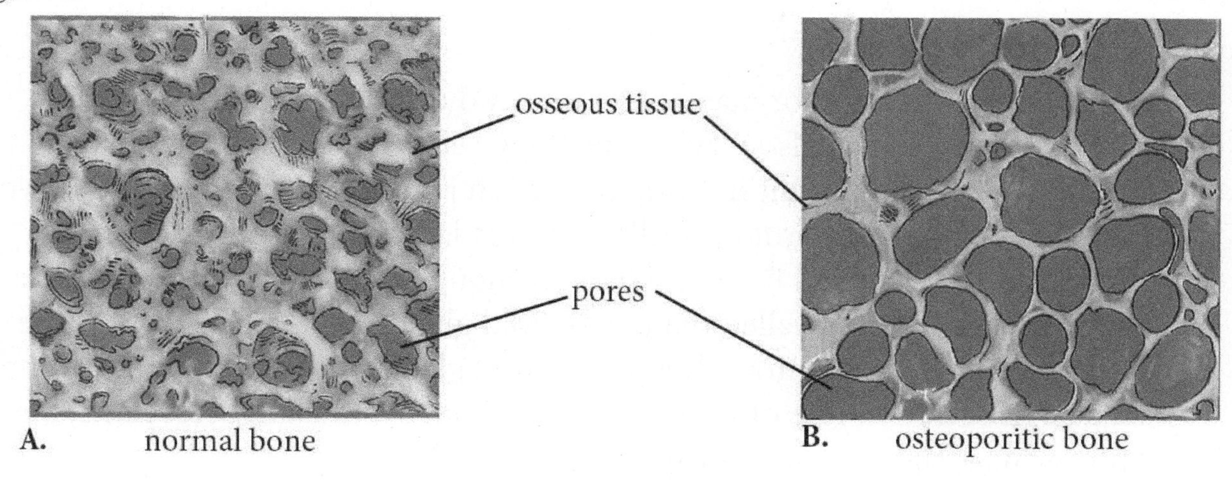

QUESTIONS:

Answer all questions with complete sentences

5. Describe at least two differences between the photos of a typical bone (figure 2A) and Suzanne's bones (figure 2B).

6. Using the information here (and not outside resources), write a definition for osteoporosis:

7. Which of the following explanations do you think is most accurate based on the information? Why?
   A. Suzanne fell which caused her hip to break.
   B. Suzanne's hip broke, causing her to fall.

Model 3: Hormones and Calcium Level Regulation

Suzanne's blood test indicates a condition called hyperparathyroidism, which means her parathyroid gland is producing an excess of parathyroid hormone (PTH). The doctor explained how this condition relates to Suzanne's broken hip and osteoporosis.

> "Calcium is necessary for many functions in the body, one being hardening of bone tissues which also serve as the body's storehouse of that mineral. Calcium is also important in cardiac function and signaling within the nervous system. The primary hormone utilized to regulate the levels of calcium circulating in our blood is parathyroid hormone, also called PTH. There is a group of small glands, called the parathyroid glands, on the back side of the thyroid gland in your throat. The parathyroid gland senses calcium levels and adjusts its secretion rate of PTH. PTH indirectly stimulates osteoclasts, the bone pruners, in order to increase blood calcium."

QUESTIONS:

8. Under what conditions is parathyroid hormone released?

9. Using the doctor's description, complete the following table. Make sure everyone in the group agrees.

Hormone	Cell type effected	Hormone's effect on Blood calcium (↑,↓, no effect)	Hormone's effect on Bone calcium (↑,↓, no effect)
Parathyroid			

10. Suzanne's mother is a visual learner and wants you to create a diagram that explains what the doctor told her about calcium in blood and bones. **On your own** (i.e., each individual within the group), use the information to create a diagram of **negative feedback** that illustrates the variable being regulated, the sensor (in this case, the sensor is also the integrator) and the effector. (Complete on a separate sheet. Limit your time to 3 minutes; this is an informal sketch.)

11. Back in your group, compare diagrams. As a group, create a new diagram of the calcium homeostasis feedback system based on the strengths of the individual diagrams. (Complete on a separate sheet. Limit your time to 5 minutes discussion and no more than 5 minutes drawing.)

12. Recall that Suzanne's blood test indicates a condition called hyperparathyroidism. With input from all group members, explain how this condition contributes to osteoporosis. *(use complete sentences)*

13. Without using outside resources, create a list of factors, excluding hyper-parathyroidism, which might result in osteoporosis. Designate which factors are within a person's control and which are not. *(Make sure your group comes up with at least three in each column)*

Factors within a Person's Control	Factors not within a Person's Control

14. Now review your answers in the table above using reliable medical websites. List the websites you consult. Star the factors in your list above that you find on the websites, and add any additional factors you find to the list.

Factors not within a Person's Control	Factors within a Person's Control

Muscle Contraction

Model 1: Anatomy of a Sarcomere

The sarcomere is the functional (contractile) unit of skeletal muscle. It is the region of a myofibril between two Z discs. Each sarcomere is approximately 2μ long.

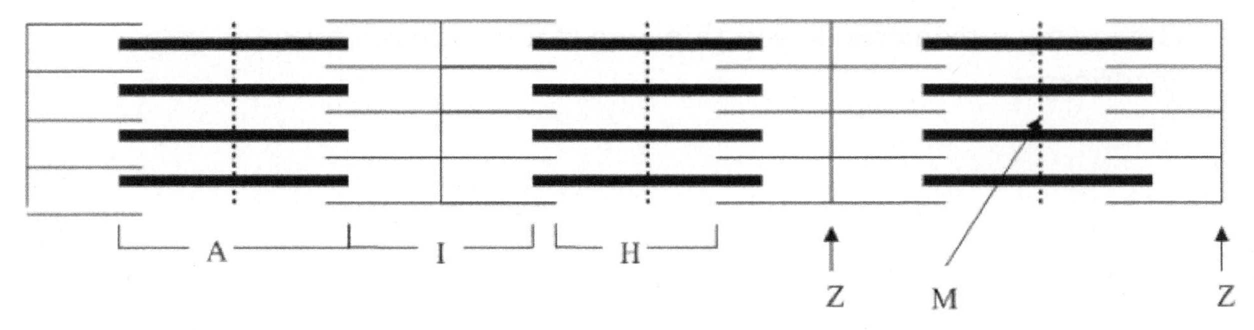

QUESTIONS:

1. Label the thick horizontal filament **THICK** filament.

2. Label the thin horizontal filament **THIN** filament.

3. How many sarcomeres are shown in the above model?

4. Based on your observations of the location of thick and thin filaments, describe each of the following:

   a) A band

   b) I band

   c) H zone

   d) Z disc

   e) M line

5. Using complete grammatically correct sentences, describe how the H zone differs from the A band.

6. How many sarcomeres do you think are in a muscle cell found in your quadriceps?

7. Do you think you would have more or fewer sarcomeres in an eye muscle?

Model 2: Comparing Relaxed and Contracted Sarcomeres

Figure 1. Relaxed sarcomeres.

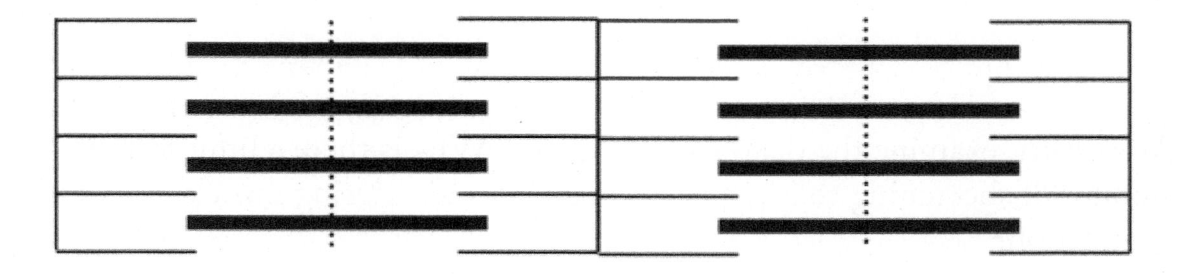

Figure 2. Contracted sarcomeres.

QUESTIONS:

8. In Figures 1 and 2 above, label the A-bands, I-bands, and H-zones. Measure and record the lengths (in mm) of these structures and the thick and thin filaments in the chart below:

Structure	Length in Relaxed Sarcomere (mm)	Length in Contracted Sarcomere (mm)	Did the length change between Figures 1 and 2? (Y/N)
Thick filament			
Thin filament			
A band			
I band			
H zone			
Sarcomere			

9. Discuss the data from the table in Question 8 with your group and describe what happens to thick and thin filaments when muscles contract:

10. As a group, examine the diagram in Model 2. Why is there a limit to the amount of shortening that can occur in a sarcomere during muscle contraction?

Model 3: Cross Sections Through a Sarcomere

Model 3 shows cross-sections of a sarcomere that show the filaments at various locations within a sarcomere.

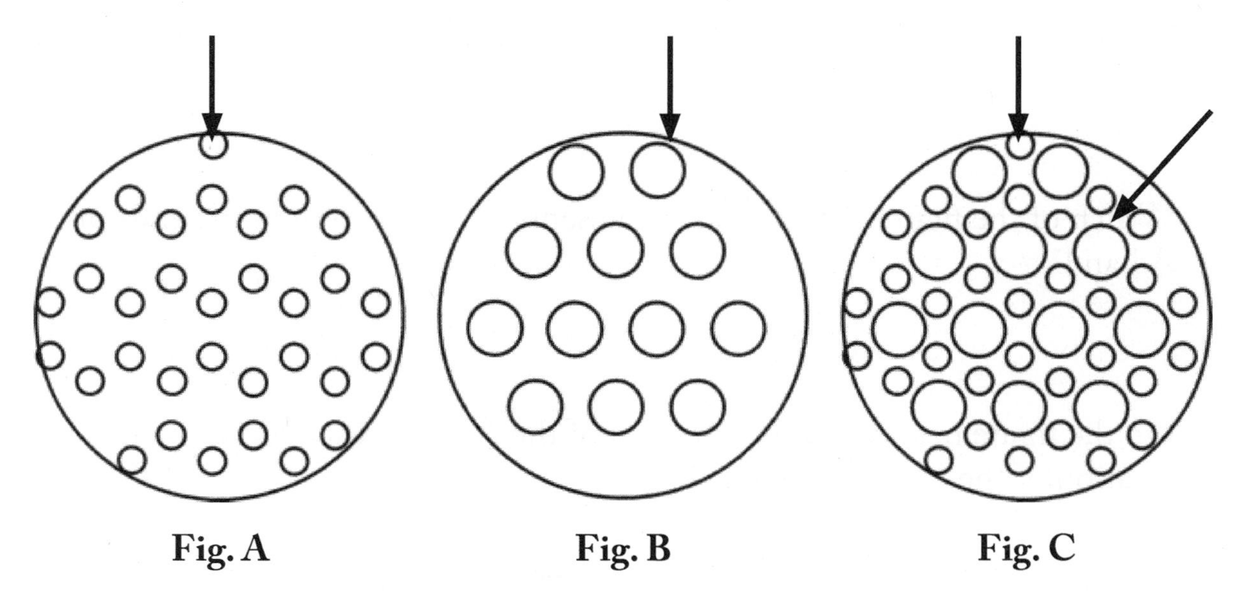

Fig. A Fig. B Fig. C

QUESTIONS:

11. Label the thick and thin filaments in Figs. A, B, and C above.

12. There are three sarcomeres shown in the diagram below.

Sarcomere 1 **Sarcomere 2** **Sarcomere 3**

a) In Sarcomere 1, identify the location within the sarcomere of the cross section indicated by Figure A in Model 3. <u>Draw a vertical line and label it **A**</u>.

b) In Sarcomere 2, identify the location within the sarcomere of the cross section indicated by Figure B in Model 3. <u>Draw a vertical line and label it **B**</u>.

c) In Sarcomere 3, identify the location within the sarcomere of the cross section indicated by Figure C in Model 3. <u>Draw a vertical line and label it **C**</u>.

13. Which of the figures (A, B, or C) represents a cross section in the H zone?

14. Which of the figures (A, B, or C) represents a cross section in the I band?

15. Which of the figures (A, B, or C) represents a cross section in the ends of the A band?

16. On the figure below, shade in the area of the A band. Then identify the location of the I band and label it.

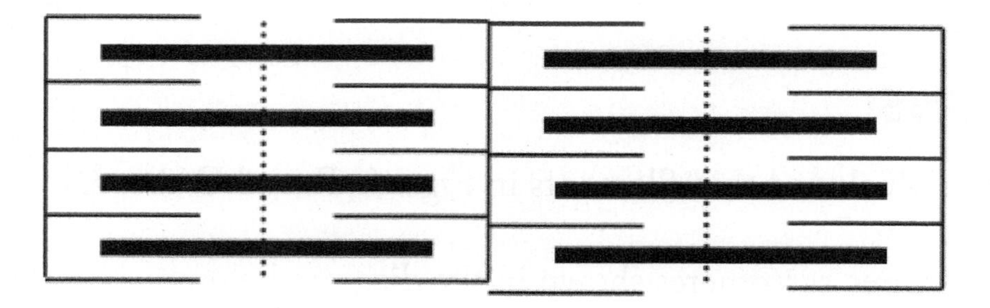

17. When viewing skeletal muscle through a microscope, you can easily see the dark and light striations of the muscle fiber. Compare the shading in the diagram in Question 16 with the photograph of the muscle fiber shown below. What forms the dark and light bands?

17. When viewing skeletal muscle through a microscope, you can easily see the dark and light striations of the muscle fiber. Compare the shading in the diagram in Question 16 with the photograph of the muscle fiber shown below. What forms the dark and light bands?

Photo courtesy of LUMEN (Loyola University Medical Education Network)

18. On the photograph above, label the A band, I band, Z disc, and a sarcomere.

19. The *sliding filament* theory is used to explain the physiology of skeletal muscle contraction. On your own, using what you have learned from this activity, write your own description of what the sliding filament theory states.

20. Next, discuss your predictions with your group members and develop a definition of the sliding filament theory with regard to thick and thin filaments. (Use grammatically correct sentences).

Heart Valves and the Cardiac Cycle

Model 1: Opening and Closing of Heart Valves

Heart valves act to keep blood flowing in one direction and prevent the back flow of blood. Heart valves open and close, somewhat like a door, because of the pressure on the two sides of the valve. When pressure on the "upstream side" is greater, the valve is open. When pressure on the "downstream side" is greater, the valve is closed.

In Model 1, Chamber A is on the "upstream side" and Chamber B is on the "downstream side" of valve 1.

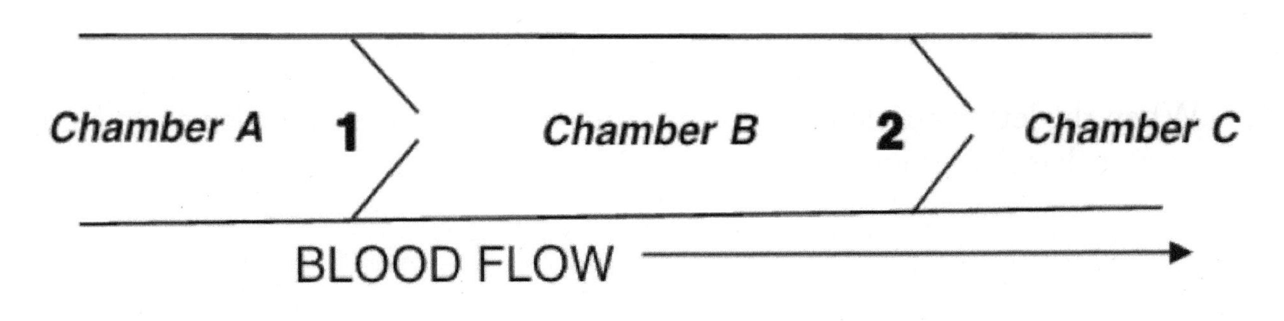

QUESTIONS:

1. When the pressure in Chamber A is greater than the pressure in Chamber B ($P_A > P_B$), is valve 1 open or closed? Explain.

2. When the pressure in Chamber C is less than the pressure in Chamber B ($P_C < P_B$), is valve 2 open or closed? Explain.

3. What relationship between pressures in Chambers A, B, and C results in both valves 1 and 2 being closed?

4. If chamber A represents the left atrium and chamber B represents the left ventricle, name the part of the heart represented by the following elements of the diagram in Model 1:

   a) Valve 1:

   b) Valve 2:

   c) Chamber C:

5. Using your answer to question 4, what do you predict would happen in a person's circulatory system if valve 2 in Model 1 did not close completely?

6. When the left ventricle pumps blood into the aorta, is left ventricular pressure greater than, less than, or equal to aortic pressure? Explain.

   a) Is the aortic (left semilunar) valve open or closed during this time? Explain.

   b) Predict if the mitral (left AV) valve is open or closed during this time. Explain.

7. When the ventricle is filling with blood, is left ventricular pressure greater than, less than, or equal to left atrial pressure? Explain.

   a) Is the mitral valve open or closed during this time? Explain.

   b) Predict if the aortic valve is open or closed during this time. Explain.

Model 2: ECG & Left Heart Pressure Changes During Two Cardiac Cycles

Model 2 shows blood pressure recorded in three different locations. Mechanical activity of the heart follows electrical activity.

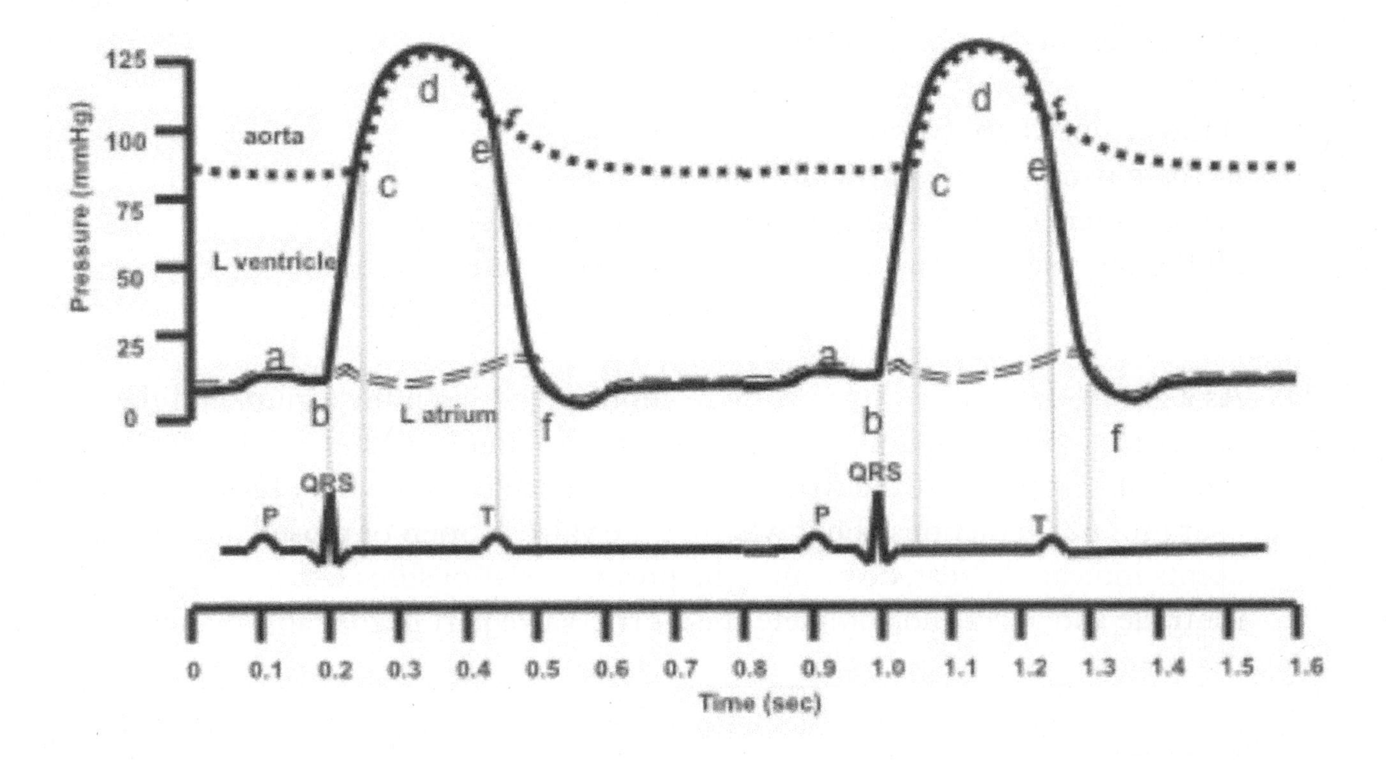

QUESTIONS:

8. Pressures are shown for which three locations in the cardiovascular system in Model 2?

   a) Which of these locations has the <u>lowest</u> pressure throughout the cardiac cycle?

   b) Which of these locations shows the <u>greatest change</u> in pressure during the cycle?

9. What are the two major phases of the cardiac cycle? Name them and briefly define each one below:

10. Using the small letters **a-f** on the pressure recordings in Model 2 indicate each of these events and intervals in a cardiac cycle:

i) the beginning of (ventricular) systole
ii) the beginning of (ventricular) diastole
iii) the interval representing systole (use two letters)
iv) the interval representing diastole (use two letters)

11. As a group, determine the pressure relationship between the atrium and the ventricle across the mitral valve at points **b, c, e,** and **f** in Model 2. Fill in those relationships for the Mitral Valve in the table below. Then fill in your group's predictions about whether the valve is **open** or **closed** at the points indicated. Next, determine the pressure relationships between the ventricle and aorta and complete the Aortic Valve part of the table below.

	Point b	**Point c**	**Point e**	**Point f**
Mitral Valve • pressure relationship	$P_{ventricle} > P_{atrium}$			$P_{atrium} > P_{ventricle}$
• valve open or closed?				
Aortic Valve • pressure relationship		$P_{ventricle} > P_{aorta}$	$P_{aorta} > P_{ventricle}$	
• valve open or closed?				

12. Use the completed table from question 11 to answer these questions about Model 2:

 a) Is the mitral valve open or closed during interval **c-e**?

 b) Is the aortic valve open or closed during interval **c-e**?

 c) Explain the importance of the conditions of these two valves during interval **c-e** in **Model 2**.

 d) Are there times during the cardiac cycle that both the mitral and aortic valves are closed? If no, why not? If yes, when and what is happening during these times?

 e) Are there times during the cardiac cycle that both the mitral and aortic valves are open? If no, why not? If yes, when and what is happening during these times?

13. When is the left ventricle pumping blood into the aorta in Model 2? *(Indicate the interval using the appropriate letters* ***a-f****)*

 a) Is the aortic valve open or closed during this interval? How does your answer to this question compare to your answer to Question 6a?

 b) Are the letters the same as the interval you listed for ventricular systole in Question 10? Explain.

14. When is the left ventricle filling with blood in Model 2? *(Indicate the interval using the appropriate letters **a-f**)*

    a) Is the mitral valve open or closed during this interval? How does your answer to this question compare to your answer to Question 7a?

    b) Are the letters the same as the interval you listed for ventricular diastole in Question 10? Explain.

15. Define each of these terms and name the small letter **(a-f)** indicating each in Model 2:

    a) **systolic** pressure in the aorta:

    b) **diastolic** pressure in the aorta:

16. Does diastolic pressure in the aorta occur during diastole or systole? Explain.

17. a) The first heart sound (lubb) immediately follows point **b** in Model 2. What causes the first heart sound?

    b) The second heart sound (dupp) immediately follows point **e** in Model 2. What causes the second heart sound?

    c) If you had only a stethoscope to examine your partner, how could you determine:

    - the beginning of systole?

    - the beginning of diastole?

    d) Is systole or diastole longer in a typical cardiac cycle? Provide evidence based on heart sounds.

Electrical Activity in the Heart

Model 1: The Excitation-Conduction System of the Heart

About 99% of the cardiac muscle fibers in one's heart contract to pump blood. The other 1% are specialized cells called "pacemaker cells" that conduct electrical impulses and can even spontaneously generate action potentials. Although pacemaker cells in different regions of the heart have different rates of generating action potentials, usually the cells with the fastest rate act as the heart's pacemaker.

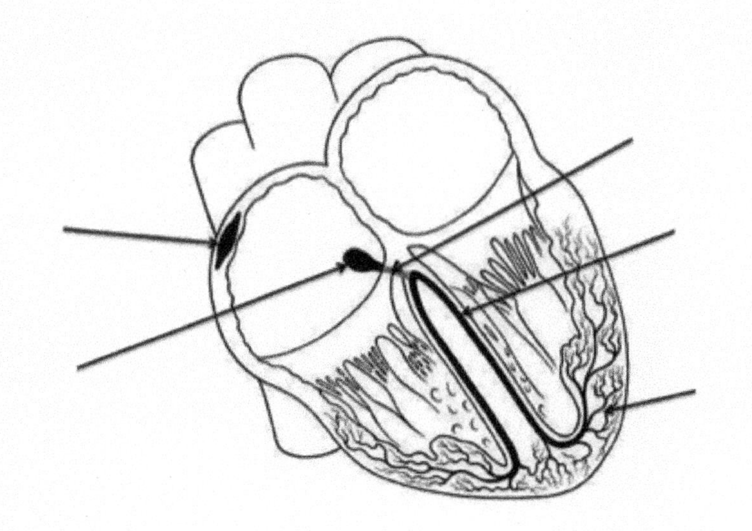

QUESTIONS:

1. Use each **bold term** described below to label the specialized excitation-conduction system of the heart in Model 1:

- The **SA node (sinoatrial or sinus node)** is located in the right atrium near the entrance of the superior vena cava. The SA node contains cells that spontaneously generate action potentials at a rate of 80-100 beats/minute.
- The **AV node (atrioventricular node)** is located at the junction between the atria and ventricles. Action potentials are propagated through the AV node very slowly. The AV node contains cells that spontaneously generate action potentials at a rate of 40-60 beats/minute.
- The **AV bundle** in the interventricular septum receives electrical activity from the AV node. This is the only pathway for electrical activity to move from the atria to the ventricles.

- Action potentials propagate from the AV bundle through the **bundle branches** to **Purkinje fibers**, which are large-diameter cells that propagate action potentials very rapidly to contractile myocardial cells throughout the ventricles. Purkinje fibers spontaneously generate action potentials at a rate of 20-40 beats/minute.

2. Before continuing, decide as a group which part of the intrinsic electrical system serves as the normal pacemaker for the heart. Explain your answer below:

Model 2: The Sequence of Electrical Excitation in the Heart

Model 2 shows the pathway of electrical excitation that passes over the heart *(darkened areas in the diagram indicate electrical excitation)*. In some areas the electrical activity spreads via the specialized conducting cardiac muscle cells. In other areas action potentials spread from one cardiac muscle fiber to the next through specializations in the cells' membranes called gap junctions.

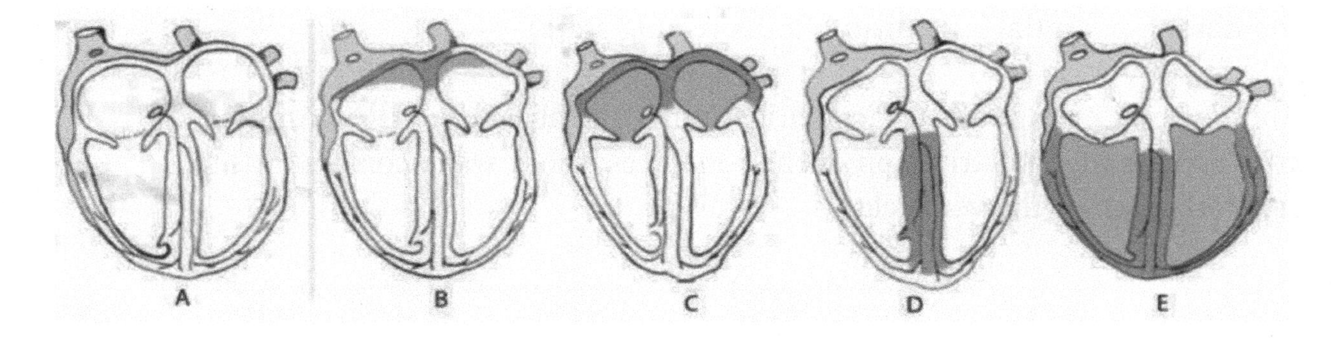

QUESTIONS:

3. Examine Model 2 and describe what is happening in each of the figures. Include the names of specific electrical system structures when they are involved. Figures A and B are described to get you started.

A: *rest (no electrical excitation)*

B: *SA node has generated an action potential which is starting to spread over the atrial muscle mass*

C:

D:

E:

Model 3: Electrocardiogram (EKG or ECG)

The electrocardiogram (ECG) shows a record of electrical activity in the heart. An ECG records the sum of all the electrical events occurring in all the cells of the heart at any point in time. Each wave represents either a **depolarization** (the membrane potential during the early part of an action potential when the membrane potential is less negative than at rest) or a **repolarization** (the membrane potential during the later stages of an action potential when the membrane potential is returning to its resting state).

The **P wave** corresponds to depolarization of the atria; the **QRS complex** corresponds to depolarization of the ventricles; the **T wave** corresponds to repolarization of the ventricles.

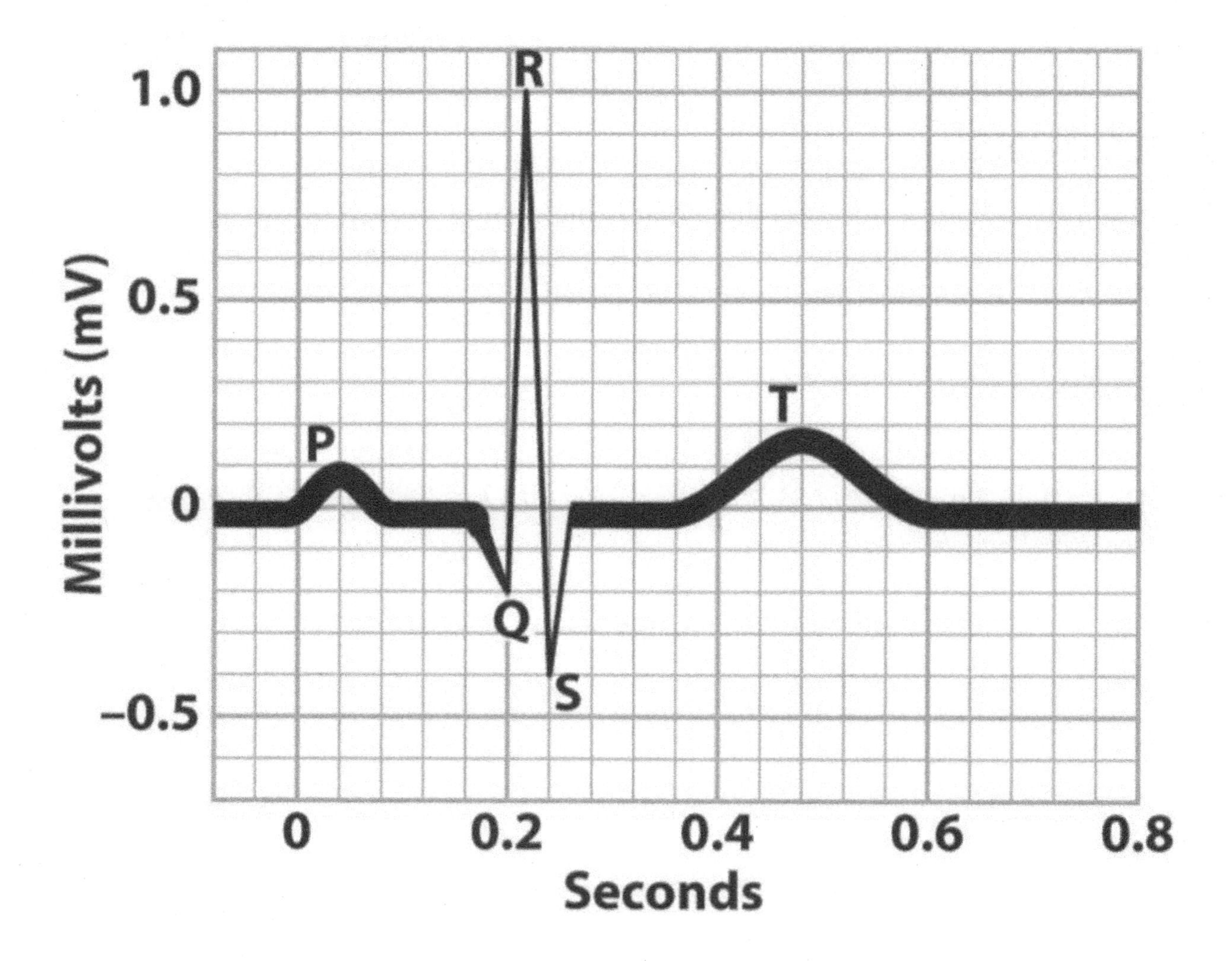

QUESTIONS:

4. Place an arrow on the ECG recording in Model 3 to indicate the SA node generating an action potential.

5. Although the ECG does not directly measure mechanical events, scientists and clinicians make assumptions about the heart's mechanical activity based on the ECG. Answer the following questions about this relationship between the ECG and the heart's mechanical activity using information from Models 1, 2 and 3:

   a) What does the P wave represent?

   Predict the mechanical event that follows the P wave.

   b) What does the QRS complex represent?

   Predict the mechanical event that follows the QRS complex.

   c) What does the T wave represent?

   Predict the mechanical event that follows the T wave.

Model 4: ECG and Selected Arrhythmias

ECGs are helpful in determining changes and irregularities (arrhythmias) in the heart's excitation-conduction system. Below are 4 ECG recordings that show a normal heart rate of 75 beats/min and three examples of normal changes in SA Node activity.

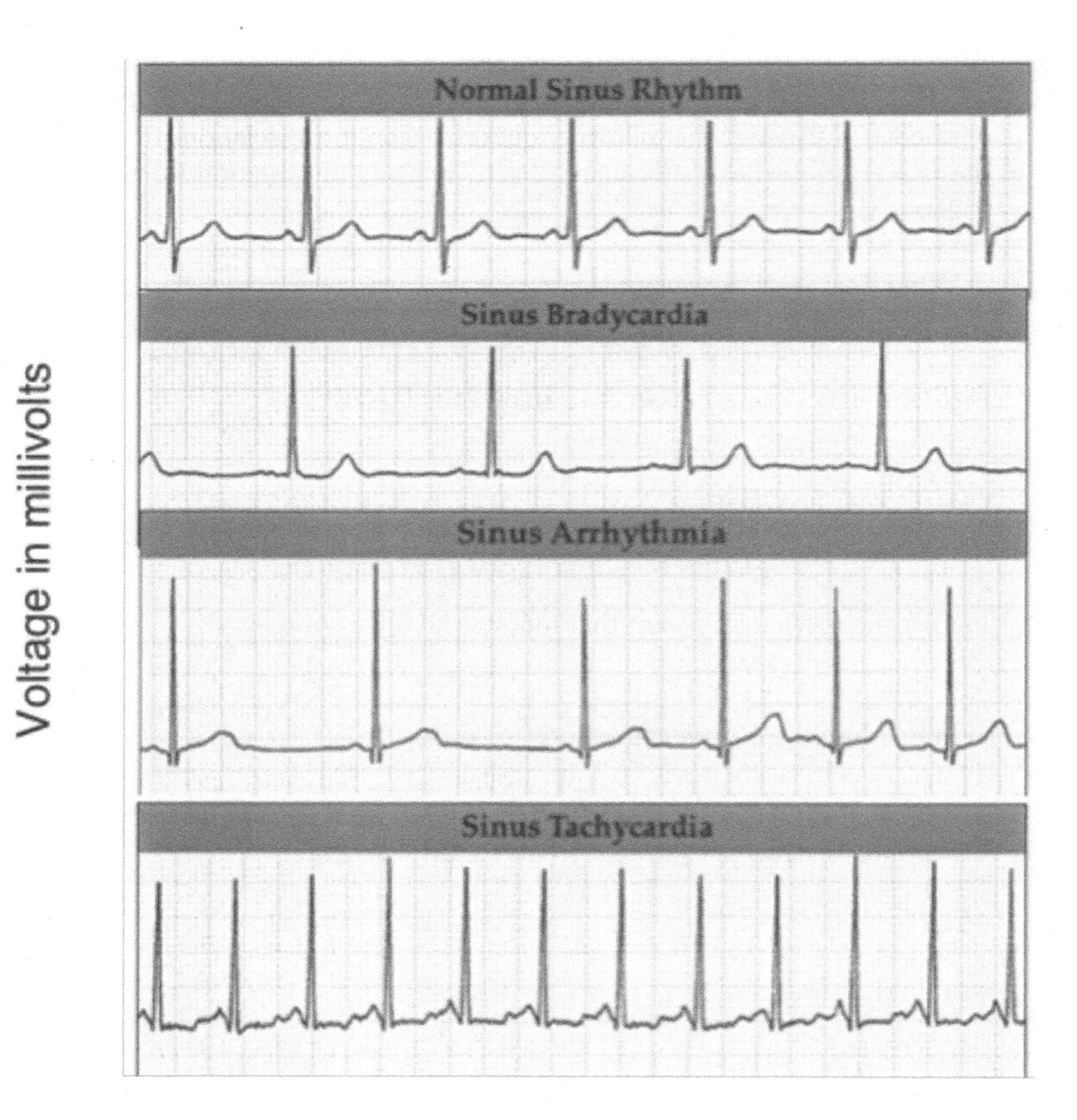

QUESTIONS:

6. Much maligned Lance Armstrong has a resting heart rate of 35 beats/minute. Which recording shown in Model 4 could be Lance's at rest? Explain your answer:

7. When Lance is pedaling up a mountainside, his heart rate is 150 beats/minute. Which recording in Model 4 could be Lance's during strenuous exercise? Explain your answer:

8. Most people have a change in heart rate associated with breathing, e.g. heart rate often increases with inhalation and decreases with exhalation.

   a) Which recording in Model 4 shows this condition?

   b) How can you determine if you have the condition described in your answer to 8a? Do you have this condition?

9. People have been known to live without atrial depolarization. What would an ECG look like in a person with this condition? Write a sentence or draw an ECG to explain your answer.

10. Given that people can live without atrial depolarization, do you think people can live without ventricular depolarization? Explain.

Model 5: Ventricular Fibrillation

Basketball player Hank Gathers collapsed and died during a college basketball game. The cause of his collapse was an irregular heartbeat. He suffered from exercise-induced ventricular tachycardia but developed ventricular fibrillation during the game. An automated external defibrillator (AED) was used to try and treat him for this condition.

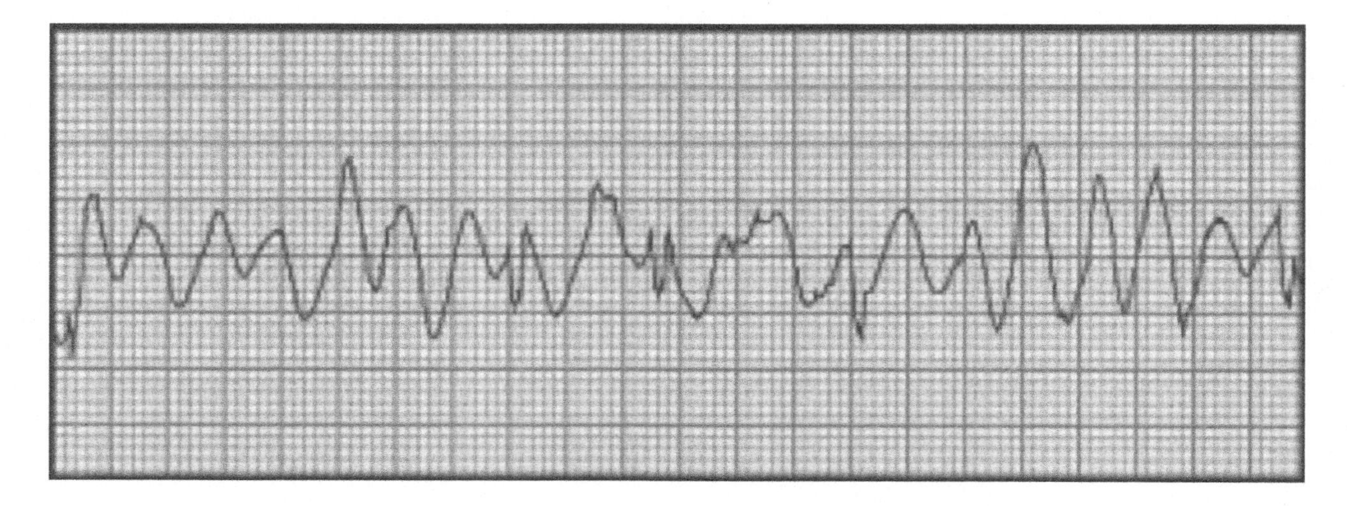

QUESTIONS:

11. Describe how this ECG of ventricular fibrillation compares to the normal ECG in Model 3.

12. What is the heart rate of the ECG shown in Model 5? Explain.

13. **Ventricular fibrillation (V-fib)** is a condition of uncoordinated contraction of the cardiac muscle of the ventricles, making them quiver rather than contract properly. People who are not health professionals usually cannot feel a pulse. Only an ECG can confirm such an arrhythmia.

    As a group, explain why this is a medical emergency that requires prompt treatment.

14. Is V-fib the same as or different from the person in question 10 who had no ventricular depolarization? Explain.

15. An **automated external defibrillator** or **AED** is a portable electronic device that automatically diagnoses the potentially life threatening cardiac arrhythmias of ventricular fibrillation and ventricular tachycardia. It is also able to treat them through defibrillation, which stops the arrhythmia, allowing the heart to reestablish an effective rhythm.

    List three places in your community that you might find an AED.

Parameters of Vascular Function

Model 1: Relationships between Pressure and Flow in a Single Vessel

The following data were collected by perfusing individual arterioles and measuring the relationship between pressure at the proximal end of the vessel and flow through the vessel. Pressure at the distal end of the vessel was maintained at 10 mmHg in all cases. The approximate experimental setup is shown in Model 1A and the data are shown in Model 1B.

Model 1A: System for collecting data shown in Model 1B

Model 1B: Data collected by varying the proximal pessure into the vessel and recording the flow rate through the vessel

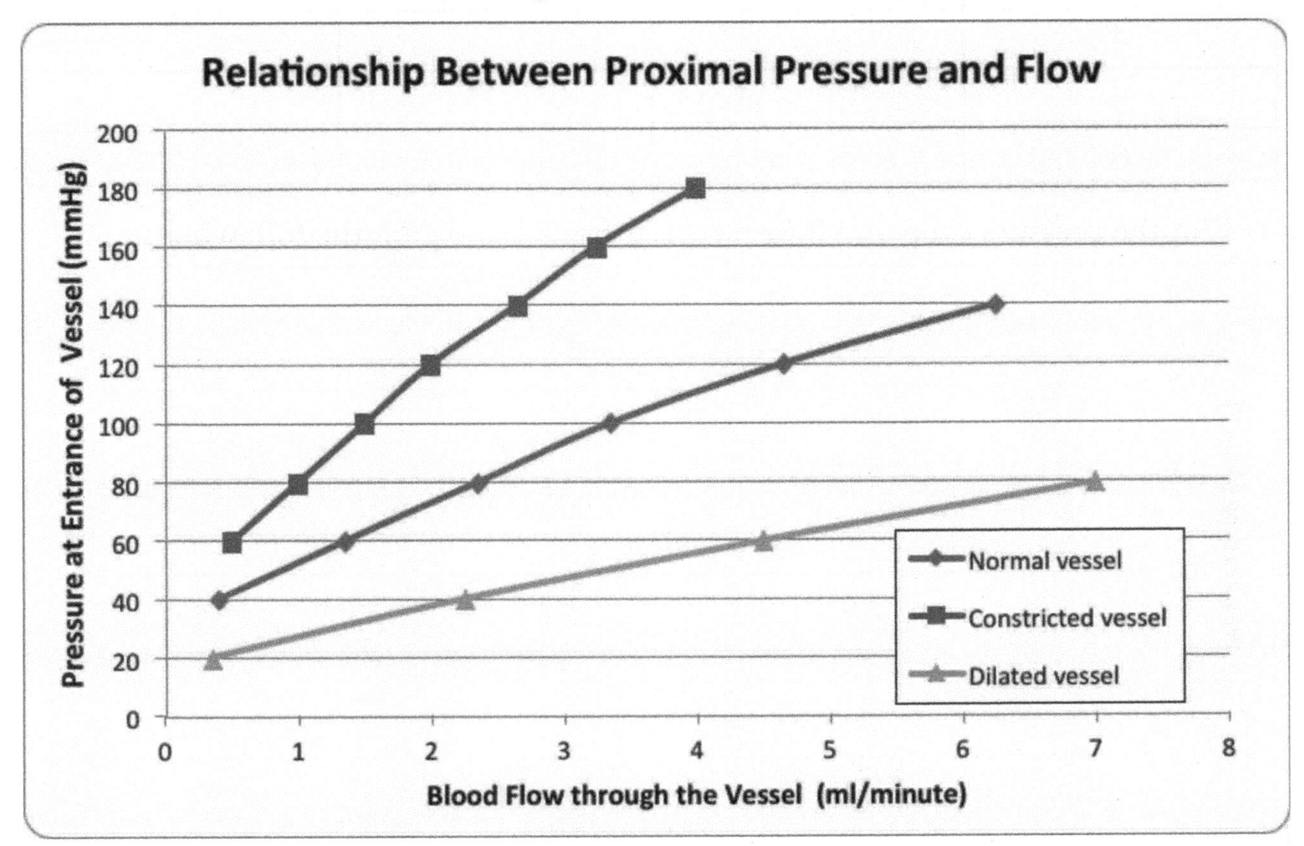

QUESTIONS:

1. What does the *proximal end* of the blood vessel refer to?

2. What does the *distal end* of the blood vessel refer to?

3. Complete Table 1 using values for the "normal" vessel.

Table 1

Blood flow rate (ml/min)	Approximate pressure at the proximal end of the vessel (mmHg)	Pressure at the distal end of the vessel	Pressure change (also knowns as pressure drop) along the length of the vessel (mmHg)
1			
2			
3			
4			
5			
6			

4. Plot the pressure drop and flow rate data from Table 1 on the following axes.

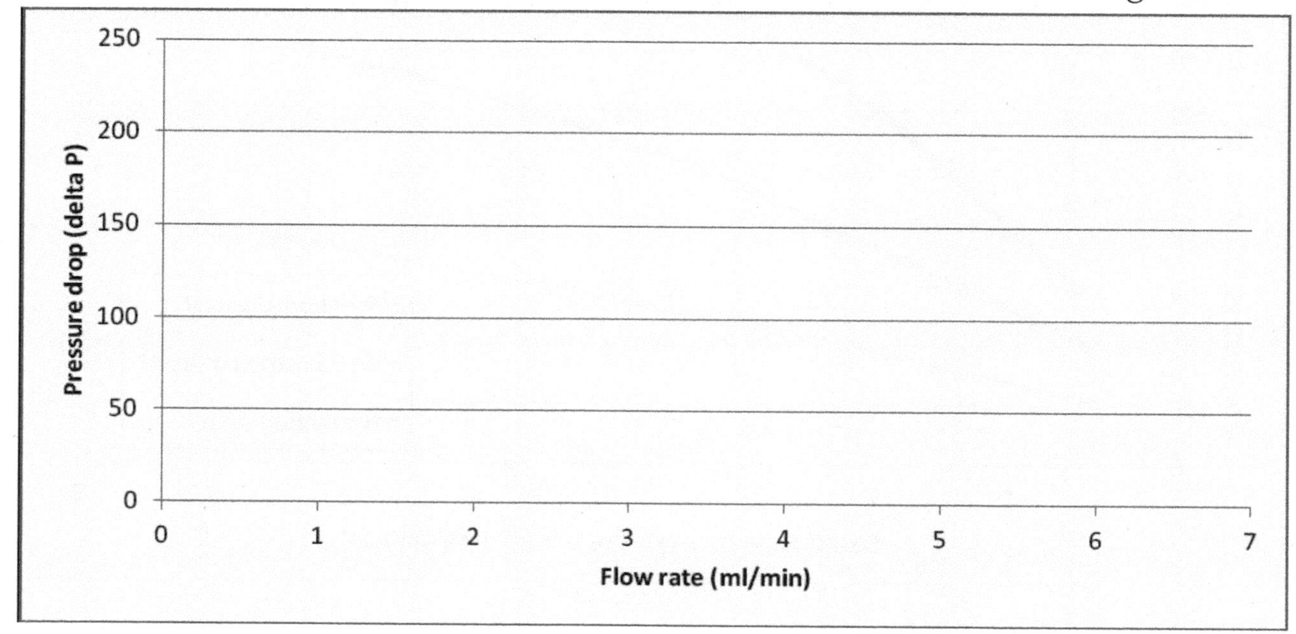

5. Although not completely linear, the relationship between pressure drop and flow rate is typically approximated by a straight line. Draw a "best fit" line through your data points.

6. Based on what you already know, write the algebraic equation for a straight line. Indicate which variables in your equation represent pressure drop and flow rate.

7. The slope of the best fit line represents the vascular resistance created by the friction between the flowing blood and the wall of the blood vessel. Assuming that the y-intercept is 0 (although hard to tell from the graph, the relationship is less linear at the very low end of flow rate), write the equation showing the relationship between pressure drop, blood flow and vascular resistance.

8. Complete the following table (Table 2): Refer to Table 1 for the intermediate steps necessary to complete the table.

Table 2

Blood flow rate (ml/min)	Approximate pressure drop across the dilated vessel	Approximate pressure drop across the constricted vessel
1		
2		
4		
6		

9. Plot the data from Table 2 on your graph in #4, including "best fit" lines.

10. Complete the following table (Table 3).

Table 3

Vessel	Vessel diameter (1 = smallest, 3 = largest)	Vascular resistance (1 = smallest, 3 = largest)
Constricted		
Normal		
Dilated		

11. What is the relationship between vessel diameter and vascular resistance?

12. What effect does increased sympathetic nervous system tone (level of activation) generally have on vascular resistance?

13. A condition called anyphylactic shock is caused by severe vasodilation secondary to an allergic reaction (such as to a bee sting). What effect would an anaphylactic reaction have on vascular resistance?

Model 2: Pressures in the Circulation

This figure shows the approximate pressures in each segment of the vasculature of a horizontal individual who has a cardiac output of 5 L/min. The solid line represents the mean pressures under normal conditions; the dotted line indicates how pressures might change under conditions of arteriolar constriction caused by the sympathetic nervous system. The left side of each region indicates the proximal end of the vessels in the region, and the right side of each region represents the pressure at the distal end of the vessels in that region. The mean (average) pressure can be approximated as the "middle" value of pressure in each region.

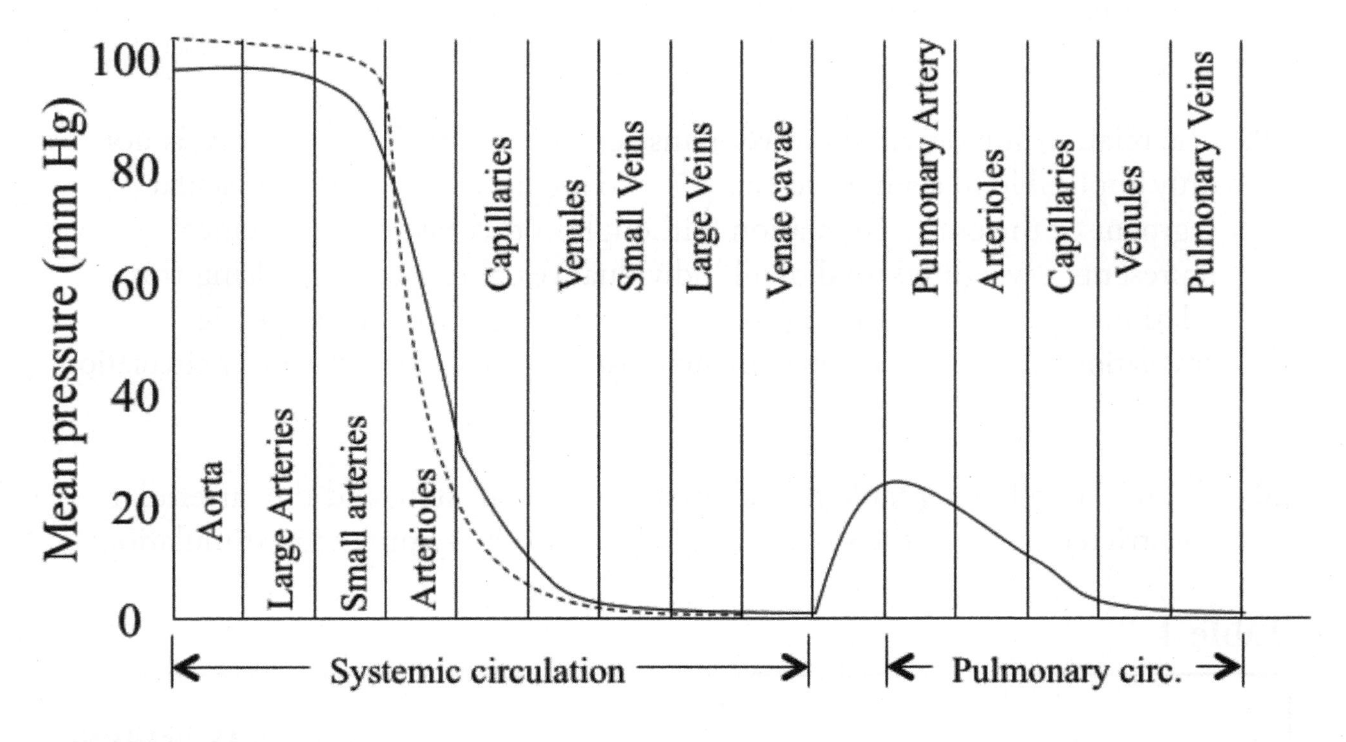

QUESTIONS:

14. How does the mean pressure compare in each subsequent order of the systemic circulation?

15. How does the mean pressure compare in each subsequent order of the pulmonary circulation?

16. How does the flow rate compare in each subsequent order of the systemic circulation? What about the pulmonary circulation?

17. What is responsible for the increase in pressure imparted to the blood between the pulmonary veins and aorta? What about the systemic and pulmonary circulations?

18. Based on your previous answers, what do you think is necessary for blood to flow through the vascular system?

19. The relationship described in your answer to question 7, $\Delta P = F \times R$, is not only applicable to a single vessel, it is also applicable to either a vascular segment or the entire circulation (although in this case, the resistance represents a weighted total of all individual vessel resistances). Using this relationship, calculate the approximate resistance across the systemic circulation. Calculate the approximate resistance across the pulmonary circulation.

20. Fill in the following table that shows the "normal" state and the "arteriolar constricted" state of the systemic vessels caused by sympathetic stimulation:

Table 4

Vessel Order	**Approximate pressure change along the vessel order**		**Approximate resistance along each vessel order**	
	Normal	Sympathetic stimulation	Normal	Sympathetic stimulation
Large arteries				
Arterioles				
Capillaries				
Venules				

21. Which vessel order is responsible for the greatest amount of resistance in the systemic circulation? In which vessel order does resistance increase the most during an increase in sympathetic tone?

22. What is responsible for the vessel order identified in question 21 having the greates change in resistance during an increase in sympathetic nervous system activity? Consider the structure of vessels in your answer.

EXTENSION QUESTIONS:

23. How do "vasodilator" drugs (used to treat hypertension) lower someone's blood pressure?

24. If the left heart only generated 60 mmHg pressure, predict how the graph of the pressure through the systemic circulation would change. You may elect to draw a new curve on Model 2 with your prediction.

25. Assume that vascular diameters are similar between the systemic and pulmonary circulations. If vascular diameter does not explain the difference in resistance between the two circulations, what else might be responsible?

26. Capillaries have smaller diameters than arterioles, but collectively have less resistance (as shown in Table 4). What might explain this?

Blood Glucose Metabolism

Model 1: The Oral Glucose Tolerance Test (OGTT)

The Oral Glucose Tolerance Test is often used to detect diabetes mellitus. The following procedure is used to conduct an OGTT:

- Patient is instructed to eat a normal diet during the days leading up to the test.
- Patient is instructed to fast (no eating or drinking) for 8 to 10 hours prior to the test (usually overnight).
- At the lab, the patient drinks a prepared glucose solution (the volume and concentration of the solution varies according to the patient's body weight).
- Measurements of the patient's blood glucose levels are taken every 30 minutes or every hour for 2 hours *(no additional food or drink is consumed during the administration of the OGTT)*. In most clinical settings, readings for blood glucose are taken at time zero and at two hours. In research studies using the OGTT, glucose levels are recorded more frequently and for longer durations of time.

The following data table shows acceptable minimum and maximum values for the OGTT.

Time (minutes)	0	30	60	90	120
Maximum Blood Glucose (mg/dL)	100	170	160	130	105
Minumum Blood Glucose (mg/dL)	80	150	140	110	85

QUESTIONS:

1. What do people consume at the start of the Oral Glucose Tolerance Test? How is the volume and concentration determined?

2. What is the time duration of the OGTT?

3. What is the range of normal values of blood glucose at the beginning (time zero) of the test?

4. What are the highest acceptable blood glucose levels during the test? At what time in the test do these values occur?

5. What is the range of normal values of blood glucose at the end of the OGTT?

6. Two individuals (Maria and Laura) both complain to their doctor of frequent thirst and frequent urination. Maria has the additional symptom of blurry vision and a "tingling sensation" in her fingers and toes. The doctor suspects diabetes mellitus, a disease in which blood glucose metabolism does not stay within normal ranges. The doctor orders both to undergo an Oral Glucose Tolerance Test to test this prediction. Below are the results of the OGTT.

Measurements recorded during OGTT

Time (minutes)	**0**	**30**	**60**	**90**	**120**
Maria: Blood Glucose (mg/dL)	90	140	150	135	110
Laura: Blood Glucose (mg/dL)	110	170	220	270	300

a) Which individual is more suspect of having diabetes mellitus?

b) Justify your answer comparing each person's OGTT data to data found in Model 1.

Model 2: Blood Glucose and Blood Insulin Levels

Insulin is a protein hormone that is secreted into the blood by the pancreas. Clinicians rarely measure insulin levels, but the procedure is regularly done in research labs studying diabetes.

The following data show blood insulin levels that correspond to the blood glucose levels in the OGTT used in Model 1.

Measurements recorded during OGTT

Time (minutes)	**0**	**30**	**60**	**90**	**120**
Maria: **Blood Glucose (mg/dL)**	90	140	150	135	110
Maria: **Blood Insulin (pmol / L)**	45	140	200	220	150
Laura: **Blood Glucose (mg/dL)**	110	170	220	270	300
Laura: **Blood Insulin (pmol / L)**	25	60	80	85	90

QUESTIONS:

7. What units are used to measure blood glucose? What units are used to measure blood insulin?

8. On your own, draw two graphs-- one showing Maria's glucose and insulin levels vs. time, and the second showing Laura's glucose and insulin levels vs. time. After all individuals have completed drawing graphs, compare them with your group members. *(Use a separate piece of paper if you choose)*

9. Which individual above (Maria or Laura) had the greater pancreatic response to the OGTT? How do you know?

10. On your own, write a grammatically correct sentence describing the relationship between blood glucose and blood insulin levels. After each individual is finished, compare sentences and, as a group, decide on the most accurate sentence.

11. Diabetics are often required to monitor their blood glucose levels to determine if/when they require a shot of insulin. Under what conditions should diabetic individuals give themselves a shot of insulin?

Model 3: Glycemic index and changes in blood glucose

Three college students (Carter, Linden, and Miriam) have their blood glucose levels measured for six hours. All three recorded what they ate for breakfast at 7:30 am, but did not record any other intake of food.

Carter's Breakfast: Orange juice, high fiber regular oatmeal, and a banana.

Linden's breakfast: Sugar soda pop, Chocolate Frosted Sugar bombs, and two cups of coffee (with sugar).

Miriam's breakfast: Bacon, eggs, and two cups of black coffee.

Student Blood Glucose Levels

Time (minutes)	7am	8am	9am	10am	11am	Noon	1pm
Carter: Blood Glucose (mg/dL)	70	140	140	80	80	80	115
Linden: Blood Glucose (mg/dL)	70	170	55	170	55	160	55
Miriam: Blood Glucose (mg/dL)	70	80	80	80	70	70	90

QUESTIONS:

12. Construct one graph that documents how each student's blood glucose levels changed over the time period shown. *(You may choose to use a separate piece of paper)*

13. How many times do you think each person ate during the 6-hour span? How do you know?

14. What types of foods are associated with the most rapid changes in blood glucose levels?

15. Glycemic Index (GI) is a numeric scale (ranging from 1 to 100) that serves as an indicator of how rapidly a food causes an increase in blood glucose levels. Carbohydrates that cause a rapid increase in blood glucose have high numbers, whereas carbohydrates that cause a gradual increase in blood glucose have lower numbers.

    a) Using the term "glycemic index," describe Carter, Linden, and Miriam's breakfasts.

    b) Which breakfast had the highest glycemic index?

    c) Which breakfast had the lowest glycemic index?

16. Using the term "glycemic index," write one or two sentences that describe the components of a healthy breakfast; one that would <u>not</u> cause a rapid change in blood glucose.

CHALLENGE QUESTIONS:

17. It is sometimes very dangerous to give someone a shot of insulin. Under what conditions should insulin *never* be administered?

18. Explain how a disease of the pancreas might cause problems with glucose metabolism:

19. Using your knowledge of medical terminology, define the term *hyperinsulinemic hypoglycemia*. What might cause this condition?

20. Many people have preconceptions about diabetes mellitus. Not all of them are correct. As a group, brainstorm what you think you know about diabetes mellitus. For example, what do you think causes diabetes mellitus? What are the long-term effects of diabetes? Can diabetes kill you? If so, what is the COD (Cause of Death)?

Determinants of Fluid Exchange (Filtration) across the Capillary Wall

Model 1: Hydrostatic Pressure

Model 1 shows a 90-foot long soaker hose. The hose is attached to a faucet at its left end and has an adjustable end cap at its right end, which affects the amount of flow out of the hose (the end cap functions like a valve). The hose is porous – it has small holes in it for water to leak across the wall of the hose. Under normal conditions, water enters the hose at 60 mmHg pressure (standard plumbing pressure in a home) and a small amount of water leaks out the right end of the hose while the rest leaks across the wall of the hose. The average pressure in each region is indicated. The pressure outside of the hose is 0 mmHg (atmospheric pressure).

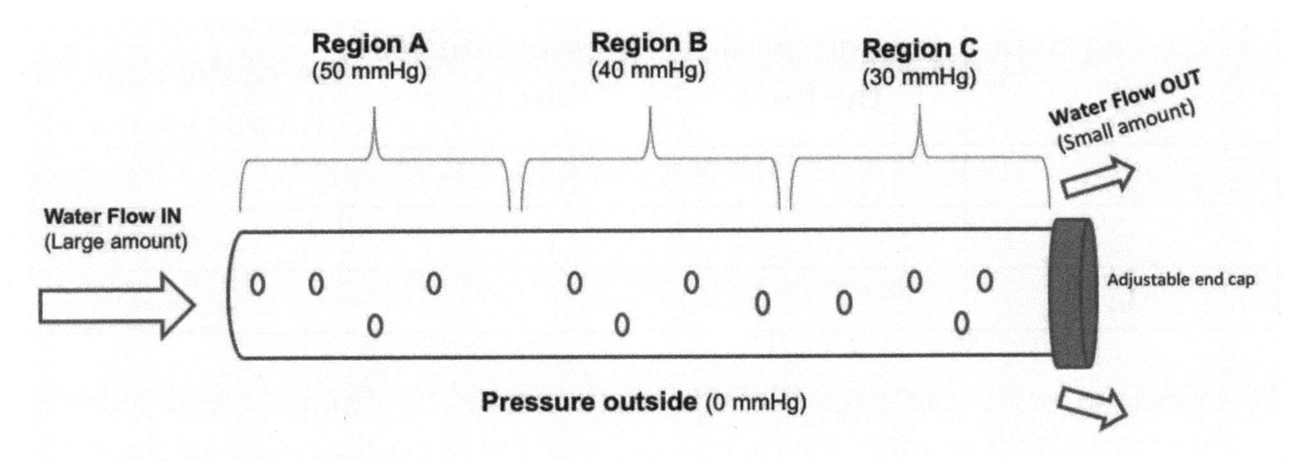

QUESTIONS:

1. In which region is the average pressure inside the hose the highest?

2. In which region is the average pressure inside the hose the lowest?

3. Predict the pressure at the end cap and explain your reasoning:

4. Based on your answers to Questions 1-3, explain how the direction of flow along the length of the hose is related to the pressure along the length of the hose. *(Hint: You might consider using a graph to show the relationship)*

Before moving on, please check your answer to Question 4 with your instructor.

5. Differences in pressure, or gradients, across the wall of the hose are calculated by subtracting the pressure outside of the house from the pressure inside the hose. For example, to find the pressure across the wall of the hose in Region A, subtract the pressure outside of the hose from the measured pressure in Region A. Using the information presented in Model 1, complete the following table:

Hose Segment	Average pressure inside the hose	Average pressure outside the hose	Pressure across the wall of the hose (inside-outside)
A			
B			
C			

6. Draw arrows on the diagram of the hose in Model 1 indicating the direction and magnitude of water flow across the wall (not along the inside) of the hose in each region Use different sized arrows to show differences in magnitude.

7. Water flows across the wall of the hose because of hydrostatic pressure (HP). On your own, write a definition of **hydrostatic pressure.**

8. Compare your answer to Question 7 with your group members. As a team, write your best definition of hydrostatic pressure.

9. If a person loosens the end cap of the hose (allowing a greater amount of water to flow out of the end of the hose), predict what will happen to the water flow across each region of the wall of the hose. Explain your reasoning based on changes in hydrostatic pressure.

Model 2: Colloid Osmotic Pressure

Model 2 shows a capillary. The large molecules, indicated by rounded squares in the diagram, are dissolved in the plasma (the non-cellular portion of blood) and are too large to pass through the intercellular clefts. The small molecules (triangles) are dissolved in the plasma, but can easily pass through the clefts. This model shows the capillary before any molecules move through the clefts.

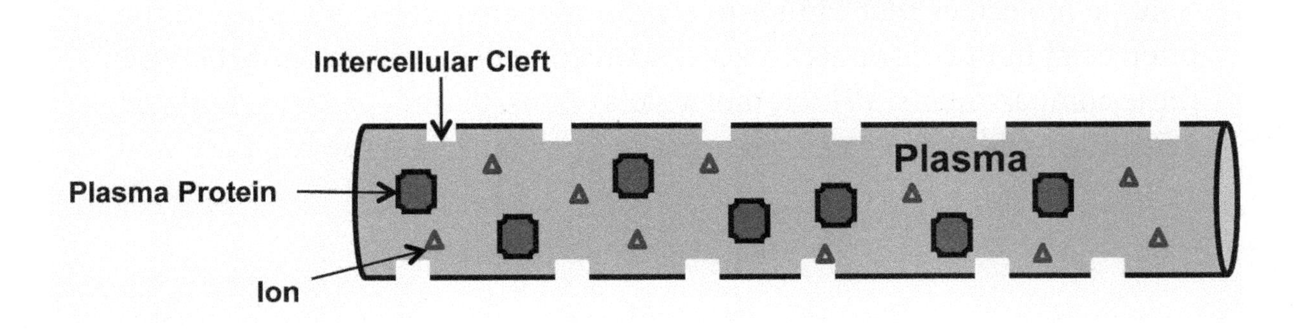

QUESTIONS:

10. a) Which type of molecules indicated in the diagram **cannot** pass through the clefts?

    b) Name at least one molecule in the human body with which you are familiar that would be included in this class:

11. a) Which type of molecules indicated in the diagram **can** pass through the clefts?

    b) Name at least one molecule in the human body with which you are familiar that would be included in this class:

12. With your group, write a working definition of **osmosis.**

Before moving on, please check your answer to Question 12 with your instructor.

13. Draw a sketch that shows how the molecules in Model 2 would be distributed after several minutes pass.

14. Soluble molecules that cannot freely diffuse between the plasma and the interstitial fluid contribute to water-attracting osmotic gradients between these compartments. Which molecule(s) (ions, plasma proteins, both, or neither) would act to create an osmotic gradient across the capillary wall?

15. Draw an arrow on the diagram in Model 2 showing the direction of the net flow of water across the capillary wall.

16. The force that attracts water into the capillary is called "colloid osmotic pressure" (COP) or "oncotic pressure." Write an individual definition of colloid osmotic pressure. As part of your definition, include any forces that contribute to COP.

17. As a team, write your best definition of colloid osmotic pressure.

18. If the concentration of large molecules, such as plasma proteins, increases in the capillary, how would this affect the magnitude of the colloid osmotic pressure? Why?

Model 3: Movement of fluid across the capillary wall (filtration)

Model 3 shows a capillary with blood flowing through it. The large molecules (indicated by the rounded squares) are too large to pass through the intercellular clefts. The small molecules (triangles) are freely permeable through the clefts and their concentration inside the vessel is the same as the concentration outside of the vessel.

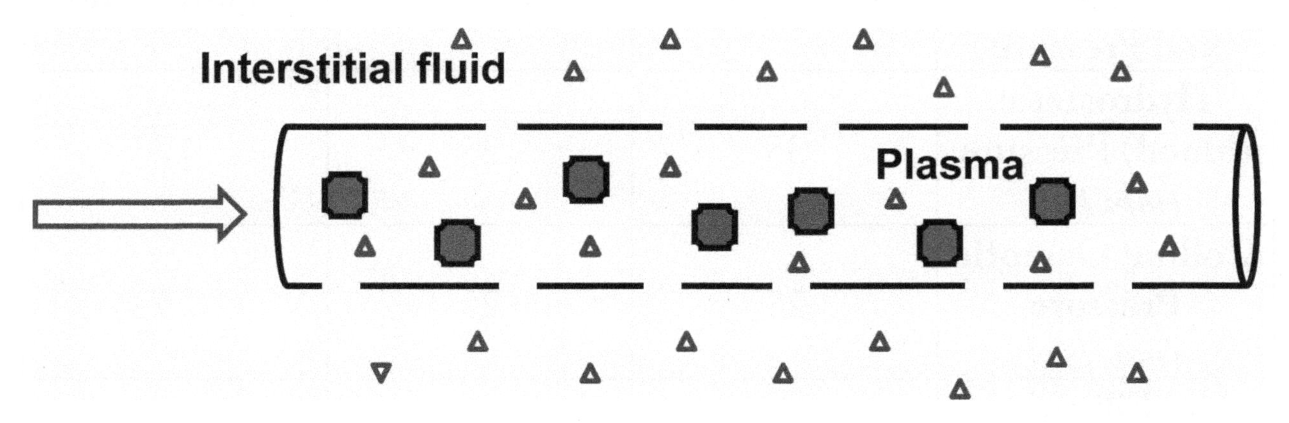

QUESTIONS:

19. Look at the image of the capillary in Model 3. The open arrow at the left indicates the location where blood enters this capillary. Label the **arterial** and **venous** ends of the capillary in the diagram above.

20. Two types of pressure were introduced in Models 1 and 2. Draw and label two arrows on the diagram in Model 3 that indicate the direction that fluid moves across the capillary wall due to these types of pressure.

21. "Blood Pressure" is the more common name for the hydrostatic pressure in blood vessels. How does blood pressure change along the length of the capillary?

22. The total pressure for fluid movement across the capillary wall is called Net Filtration Pressure (NFP). Using your answer from Question 20, write an equation for determining Net Filtration Pressure:

23. The following table shows representative values of a standard systemic capillary in the body. Complete the table using the equation you developed in question #22 and the arrows you added to the diagram in Model 3.

Capillary Region

Measures of Fluid Movement and Pressure	**Arterial End**	**Middle**	**Venous End**
Hydrostatic (blood) Pressure *(mm Hg)*	35	25	17
Colloid Osmotic Pressure *(mm Hg)*	25	25	25
Net Filtration Pressure *(mm Hg)*			
Direction of Fluid Movement *(into capillary, out of capillary, or no movement)*			

24. The magnitude of the net filtration pressure at each point helps determine how much fluid moves across the wall of the capillary. Based on your calculations, is there a net flow of fluid into the capillary, out of the capillary, or no net change in fluid moving across the capillary wall? Explain.

Extension Questions:

25. Predict how fluid movement across the capillary wall would be affected by each of the following conditions:

    a) Hypoalbuminemia:

    b) Increased venous pressure (i.e. venous congestion):

26. Some conditions, such as allergies and inflammation, can make capillaries permeable to large molecules as well. How would this affect net filtration pressure and the amount of fluid moving across the capillary wall?

27. Mannitol is a compound that is administered to patients under certain conditions because it is filtered into the nephrons of the kidney, but not reabsorbed back into the blood. The mannitol molecules end up in the bladder, where nephrons have some similarities to capillaries.

    a) What effect(s) would mannitol have in the nephrons of the kidney?

    b) What do you think mannitol is used for?

Determinants of Blood Oxygen Content

Model 1: Oxygen Binding to a Single Hemoglobin Molecule

Hemoglobin (Hb) is a large molecule consisting of four peptide chains. Each peptide contains a heme group with an iron in the center. An oxygen molecule has the ability to bind to each heme group under the appropriate conditions. The number of oxygen molecules bound determines its saturation.

Saturated Hemoglobin Molecule

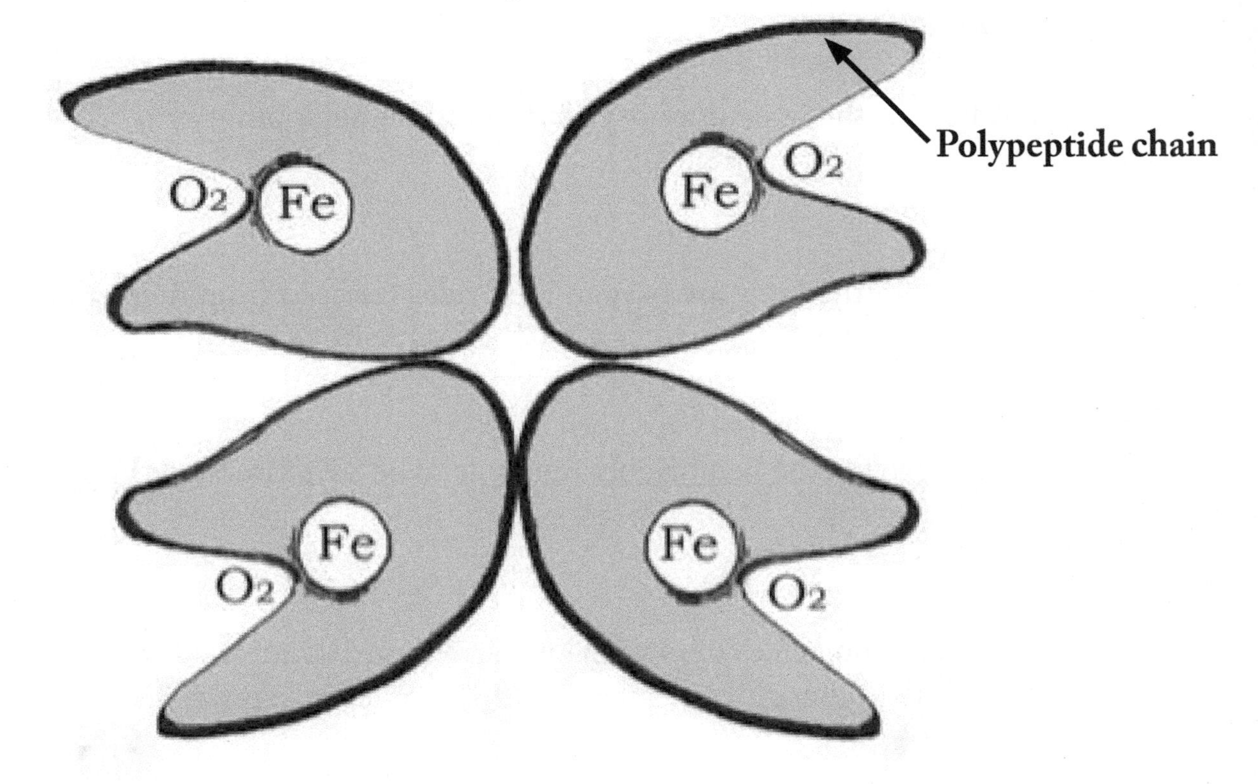

QUESTIONS:

1. When a hemoglobin molecule is completely saturated, how many oxygen molecules are attached?

2. When a hemoglobin molecule is 75% saturated, how many oxygen molecules are attached?

3. Assuming you started with the molecule in Model 1, how many molecules of oxygen would need to be released in order for the hemoglobin to be 75% saturated?

4. The number of oxygen molecules bound to hemoglobin tends to fluctuate up and down. *(Discuss the following questions with your group, and then write your answer)*

   a) Where in the body does oxygen bind to hemoglobin?

   b) Where in the body are oxygen molecules released from hemoglobin?

5. Is it possible for <u>one hemoglobin molecule</u> to be 82% saturated? Why or why not?

6. How many molecules of oxygen are carried on 100 molecules of hemoglobin when all the hemoglobin molecules are completely saturated?

7. If after passing through a capillary bed, those 100 molecules of hemoglobin become 75% saturated with oxygen, how many molecules of oxygen were unloaded (released)?

Model 2: Hemoglobin (Hb) Values and Their Effect on Oxygen Carrying Capacity

Hemoglobin is found in very high numbers within red blood cells, and there are lots of red blood cells in our blood. Because dealing with such large numbers can be challenging, hemoglobin is typically measured by mass and oxygen is measured by volume. The following table provides information about common parameters used in calculating blood oxygen levels.

Parameter	Normal Value
Number of hemoglobin molecules per red blood cell	250 million molecules
Number of red blood cells per deciliter (dL) of blood (There are 10 deciliters in 1 liter of blood)	500 billion cells
Mass of hemoglobin in a deciliter (dL) of blood	15 grams (g)
Amount of oxygen that 1 gram of hemoglobin can hold (carry) when 100% saturated	1.34 milliliters (mL)

QUESTIONS:

8. How many molecules of oxygen can one red blood cell carry?

9. Is it possible for the total hemoglobin inside a red blood cell to be 82% saturated? Explain your answer.

10. Set up a mathematical equation that illustrates how to calculate the number of oxygen molecules that could be carried in 1 dL of blood when it is fully saturated with oxygen (you do not have to solve your calculation):

11. Calculate the amount (in mL) of oxygen that could be carried in 1 dL of blood when the hemoglobin is 100% saturated:

12. Calculate the amount (in mL) of oxygen carried in 1 dL of blood when the hemoglobin is 50% saturated:

Model 3: Oxyhemoglobin Dissociation Curve

A major determinant of how much oxygen is found in blood is the saturation of hemoglobin (which determines the amount of "bound" oxygen). This saturation level is mainly determined by the amount of oxygen dissolved in the blood plasma, because the dissolved and bound oxygen are in chemical equilibrium. The amount of dissolved oxygen is indirectly measured as PO_2, the partial pressure of oxygen in plasma. When the PO_2 increases, saturation typically increases as well. This relationship is illustrated by the oxyhemoglobin dissociation curve, as shown below.

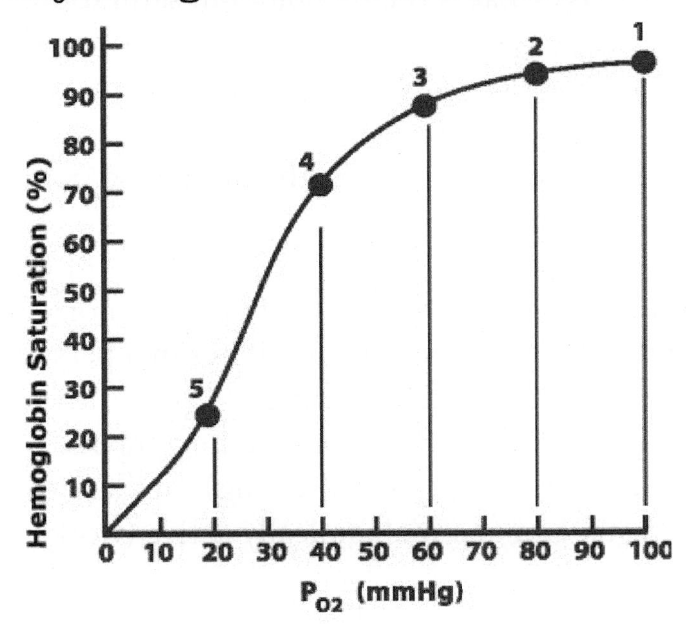

QUESTIONS:

13. What is the label on the X-axis?

14. What are the units specified on the X-axis?

15. What is the label on the Y-axis?

16. What are the units specified on the Y-axis?
17. As partial pressure of oxygen (PO_2) increases, what happens to hemoglobin saturation?

18. At PO_2 of 100 mm Hg, what is the hemoglobin saturation?

19. At PO_2 of 20 mm Hg, what is the hemoglobin saturation?

20. Answer questions a, b, and c assuming the numbered points on the graph in Model 3 represent:

    1: Blood at the distal end of alveolar capillaries in a normal individual.
    2: Blood in the distal end of alveolar capillaries in a person with slightly impaired oxygen exchange in their lungs.
    3: Blood in the middle of a capillary of systemic organs.
    4: Blood at the distal end of a typical (resting) capillary of systemic organs.
    5: Blood at the distal end of a metabolically active (working) capillary.

    a) What is the "distal end" of a capillary?

    b) Where are the distal ends of alveolar capillaries located?

    c) Where might you find the distal end of a metabolically active (working) capillary?

21. Complete the following table. Assume a "normal" individual with a Hb concentration of 15 g/dL. There is space for your calculations after the table.

Vessel	PO_2 (mmHg)	Hb Saturation (%)	Oxygen content (mL/dL)
Pulmonary vein			
Distal end of resting capillaries			
Distal end of working capillaries			
Vena Cava	*40*		
Pulmonary Artery			

22. Using the results of your calculations in Question #21, complete the following table. *(Assume all references to capillaries refer to the distal end of the capillary)*

Relationship	**Difference (number)**
PO_2 difference between the pulmonary and resting capillaries	
PO_2 difference between the pulmonary and working capillaries	
Oxygen content difference between the pulmonary and resting capillaries	
Oxygen content difference between the pulmonary and working capillaries	

23. When a person exercises, what happens to the oxygen saturation of hemoglobin in capillaries leaving muscle tissue? Discuss with your group and write one answer.

24. Does a change in PO_2 always correspond to a proportional change in oxygen content? Explain why or why not:

Extension Questions:

25. There is an oxygen bar in town where you can pay to breathe air with higher than normal oxygen content. The owners claim it will improve your alertness and ability to function. Using data from the graph in Model 2, discuss with your group whether you think this statement is true or not. Explain your answer in complete sentences below:

26. A person with anemia has a decreased concentration of hemoglobin in the blood.

    a) Would this affect the person's oxyhemoglobin dissociation curve? Explain:

    b) Would it affect their blood oxygen content? Explain:

27. Describe a specific situation or circulatory location that could be represented by point 3 in Model 3:

Action Potential

Model 1: Alteration of Resting Membrane Potential

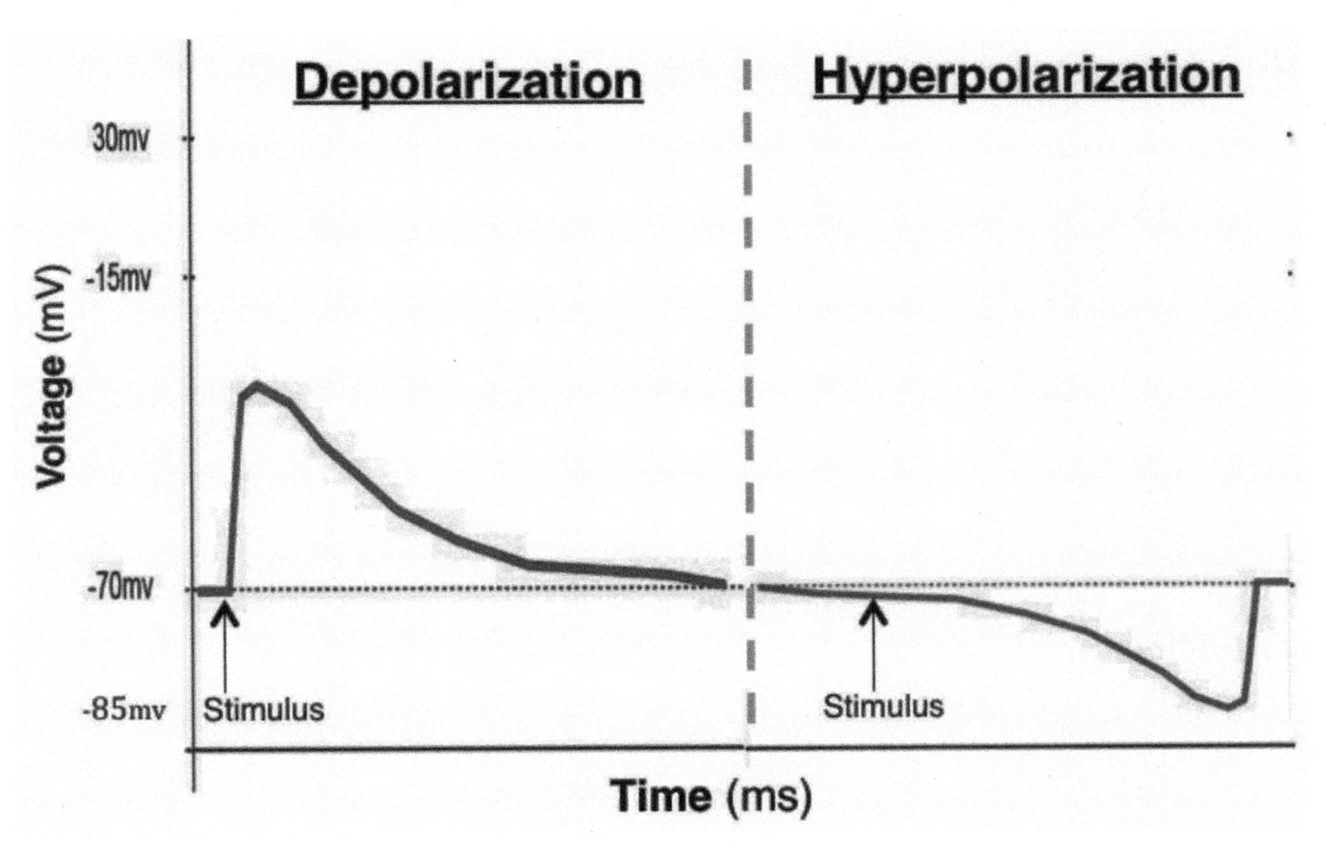

QUESTIONS:

Resting Membrane Potential (RMP) is the potential difference that exists across a membrane of an unstimulated cell (i.e. cell at rest). What is the RMP for a typical neuron shown in model 1?

1. **Depolarization** and **hyperpolarization** are two types of alterations in resting membrane potentials. Based on the model, determine if the following membrane values are depolarized or hyperpolarized. The manager should have each member of the team take turns predicting the RMP change.

    a) 0 mV =

    b) -100 mV =

    c) -71 mV =

    d) -70 mV =

    e) -69 mV =

    f) +30 mV =

2. Voltage is a recording of the inside charge compared to the outside charge. Discuss as a group the meaning of depolarization and hyperpolarization, then write a grammatically complete sentence to define each RMP alteration:

   a) Depolarization:

   b) Hyperpolarization:

Model 2: Ion Movement Across Neuronal Membrane

Neurons use changes in membrane potential as communication signals (nerve impulses). Resting membrane potential changes are due to ions crossing the plasma membrane through specific ion channels.

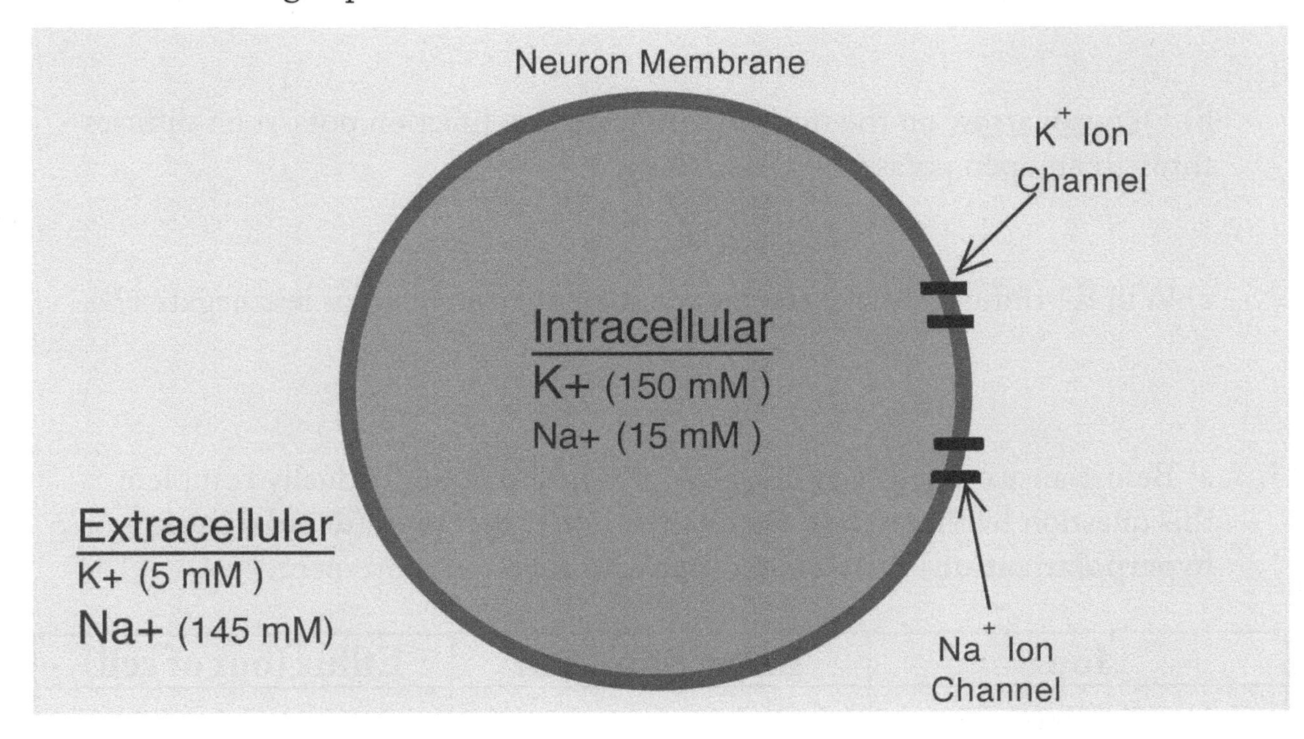

QUESTIONS:

3. Examine the location of **sodium ions** in Model 2 and answer the following questions:

   a) Is sodium more concentrated on the inside or outside of the cell membrane?

   b) <u>Draw an arrow</u> on the model to indicate the direction sodium diffuses through an open sodium ion-channel.

   c) Will Na+ diffusion make the membrane potential more or less negative?

4. Examine the location of **potassium ions** in Model 2 and answer the following questions:

   a) Is potassium more concentrated on the inside or outside of the cell membrane?

   b) Draw an arrow on the model to indicate the direction potassium diffuses through an open potassium ion-channel.

   c) Will K+ diffusion make the membrane potential more or less negative?

5. a) Before answering as a group, each person should individually complete this question by themselves. Predict the RMP alteration (**depolarization** or **hyperpolarization**) that would occur with each ion movement.

Ion	Influx (into cell)	Efflux (out of cell)
Anion (-)		
Cation (+)		

   b) Within your group, discuss and compare your individual answers to the above question. The Manager should make sure that each individual is sharing their results. Then reach a group consensus of one best answer.

6. In a resting neuron there is greater K+ permeability then Na+ permeability. What would happen to the RMP if sodium permeability increased to the same level as potassium?

Model 3: Action Potential

Model 3 shows a typical neuron action potential, which is the method by which neurons communicate. **Action potentials** are brief reversals of membrane potential in response to a stimulus. Action potentials travel along the axon to the axon terminal to stimulate neurotransmitter release.

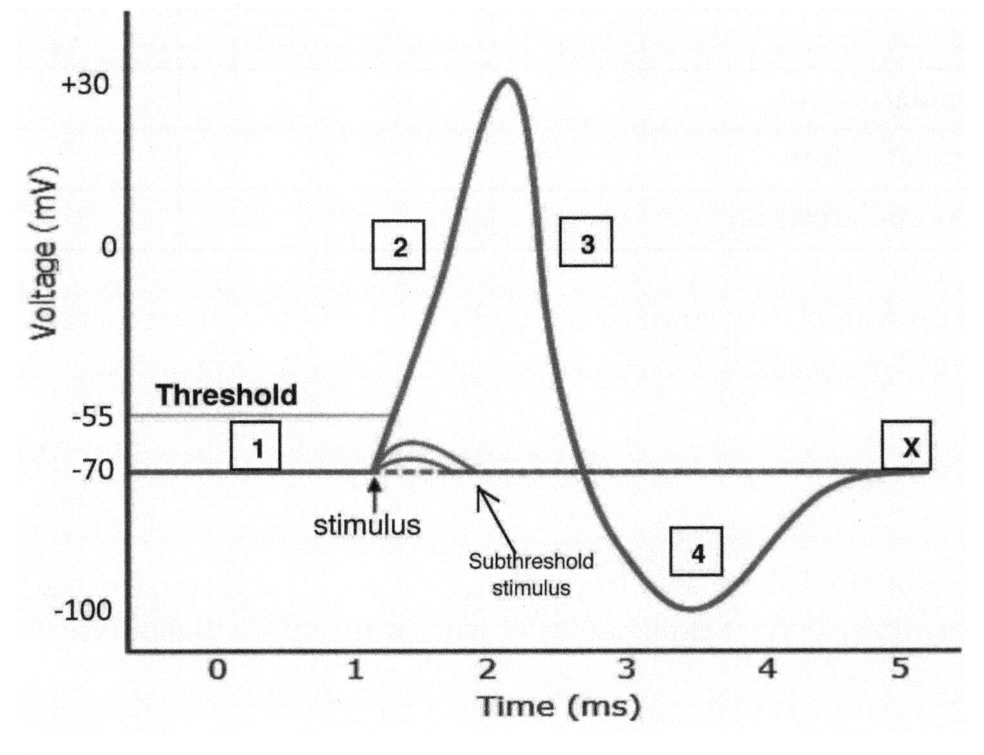

QUESTIONS:

7. Examine Model 3 to determine the voltage range for the four phases of the action potential:

   1) Resting:

   2) Depolarization:

   3) Repolarization:

   4) Hyperpolarization:

8. What is the name of the phase marked X?

9. Action potentials occur due to the opening and closing of sodium and potassium gated ion channels. As a group, predict if these Na and K voltage-gated ion channels are **open** or **closed** in each phase. Be sure that each team member provides input.

#	Action Potential Phase	Sodium (Na+)	Potassium (K+)
1	Resting		
2	Depolarization		
3	Repolarization		
4	Hyperpolarization		

10. Based on Model 3, what must be reached in order to initiate an action potential? *(be specific)*

11. Action potentials have an all-or-none principle. As a group, write a grammatically correct explanation of what is meant by this phrase.

12. Although stimulus originates at the dendrite and travels through the cell body, an action potential does not start until it reaches the axon hillock. Why do you think action potentials do not occur on dendrites (even with a very strong stimulus above threshold), and why do action potentials not start until the axon hillock?

REVIEW QUESTIONS:

13. When a Potassium voltage-gated channel opens in a neuron at rest, what will happen to RMP? (choose from: *no change, depolarization, repolarization, or hyperpolarization*)

14. (*Fill in the blank:*) A drug that opens sodium channels in a motor neuron would ____________ the membrane. Predict how this drug would affect the probability of generating an action potential:

15. Novocaine and lidocaine are two local anesthetic drugs used in dentistry that block nerves from sending action potentials to the brain. Describe possible mechanisms of action for these anesthetic drugs:

16. The ability of the heart to contract (beat) is due to an action potential signal. **Hypokalemia** and **hyperkalemia** (low and high potassium blood plasma levels) can be deadly.

    a) How do hypokalemia and hyperkalemia alter the cardiac cell membrane potential and the ability of heart muscle to contract?

    b) Which situation is more deadly--hypokalemia or hyperkalemia? Why?

The Menstrual Cycle

The **menstrual cycle** includes changes observed in the ovary, in plasma levels of reproductive hormones, and in the uterus. This activity looks at each of these cycles and the relationships among them.

Model 1: Ovarian Events

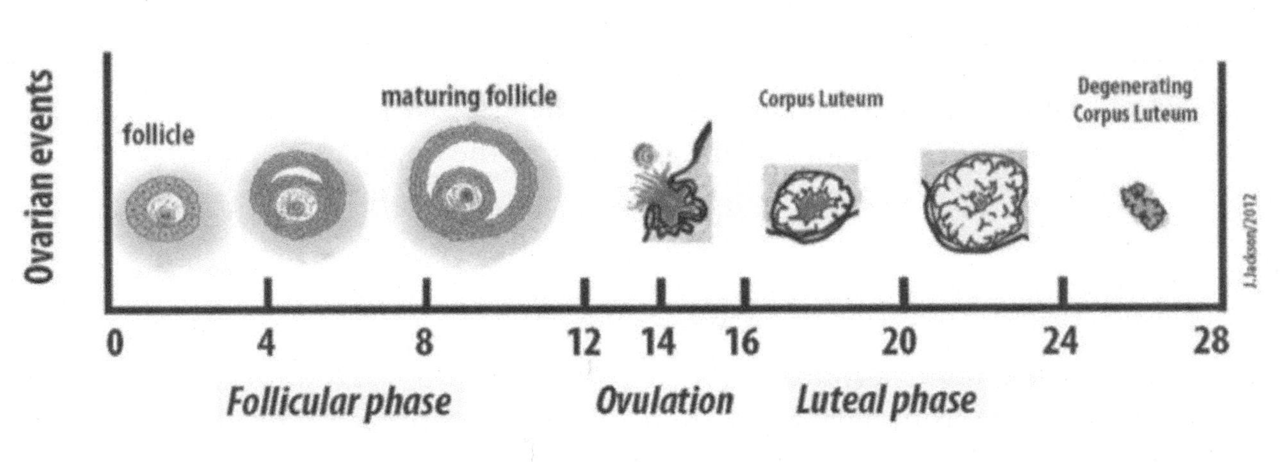

QUESTIONS:

1. How many days are shown for this typical ovarian cycle?

2. Look closely at the follicles represented between days 0 and 4. At the center of the follicle is the oocyte (developing egg), and surrounding it are the follicular cells. In the space below, re-draw this whole structure and label both the <u>oocyte</u> and the <u>follicular cells</u>.

3. What changes can be seen in the follicle between days 0 and 12?

4. The ovarian cycle has two phases (follicular and luteal) and one event (ovulation). Complete the table below with information from Model 1:

Ovarian Phase Name	Days of Cycle	Ovarian Changes
	0-13	
Ovulation	14	
	15-28	

5. After ovulation, the follicular cells become a structure called the corpus luteum. What is ovulation? What happens to the oocyte?

6. The corpus luteum secretes estrogen and progesterone. Examine the size of the corpus luteum from days 16 to 28 in Model 1 and predict how changes in the size of the corpus luteum affect its endocrine function.

Model 2: Ovarian Events + Plasma Levels of Sex Steroid Hormones

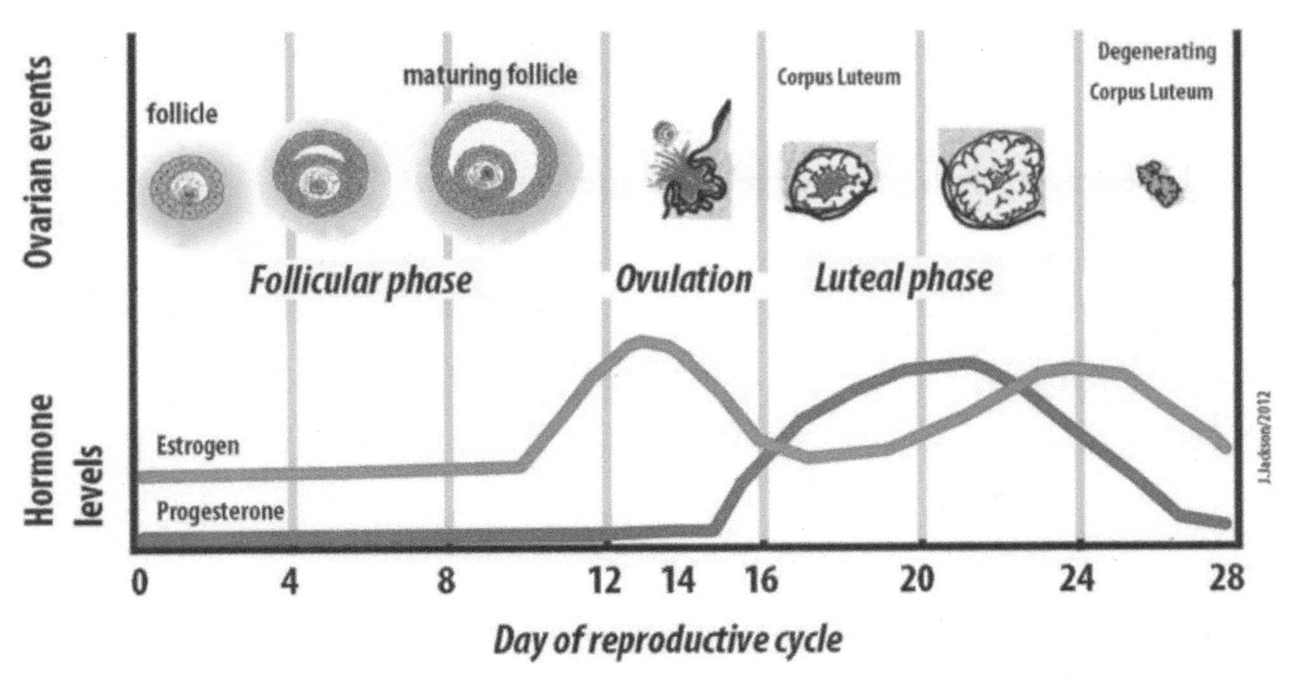

QUESTIONS:

7. Name the two sex steroid hormones shown in Model 2.

8. Name the ovarian phase(s) during which these hormones show changes in concentration:

   a) Estrogen

   b) Progesterone

9. Name the ovarian phase during which each hormone reaches its peak:

   a) Estrogen

   b) Progesterone

10. Why do you think plasma levels of both sex steroid hormones decrease near the end of the cycle?

11. Do estrogen levels peak before, at, or after ovulation?

Model 3: Ovarian Events and Plasma Levels of Gonadotropins and Sex Steroid Hormones

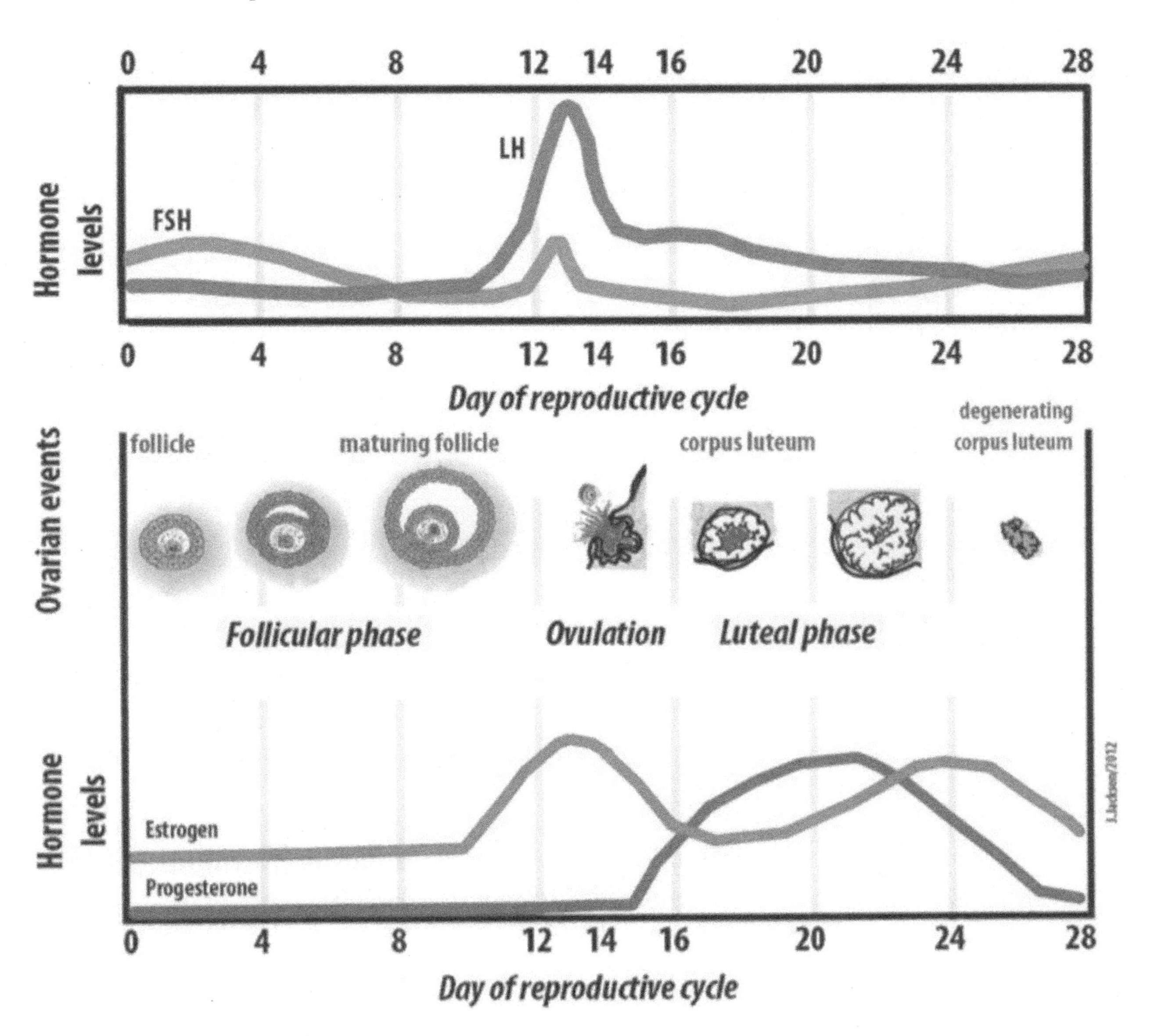

QUESTIONS:

12. The anterior pituitary secretes two gonadotropins (hormones that target the gonads). Name them, including both the abbreviated and long versions of their names.

13. Using the diagram below, show the location of the anterior pituitary and the ovaries. How do gonadotropin hormones move from the anterior pituitary to the ovaries?

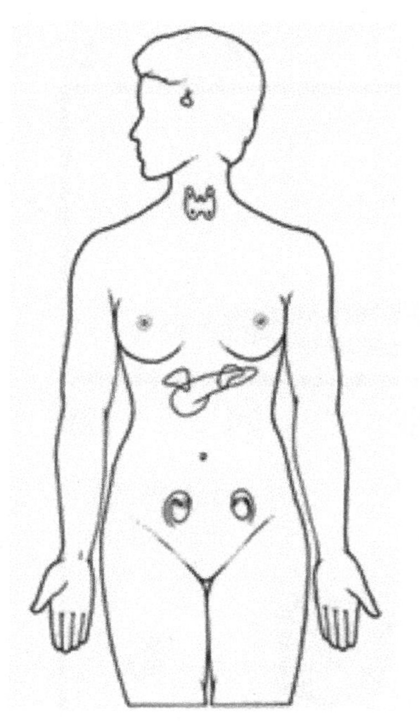

14. Plasma levels of gonadotropins peak on day 13. What ovarian <u>event</u> immediately follows the surge in LH?

15. What do you think is the function of **Follicle Stimulating Hormone (FSH)**? *(Hint: look closely at the relationships illustrated in Model 3)*

16. What do you think is the function of **Luteinizing Hormone (LH)**? *(Hint: look closely at the relationships illustrated in Model 3)*

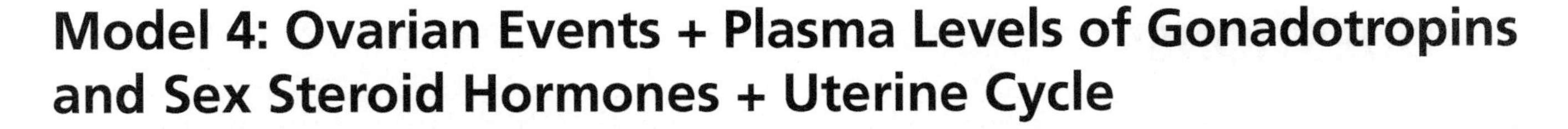

Model 4: Ovarian Events + Plasma Levels of Gonadotropins and Sex Steroid Hormones + Uterine Cycle

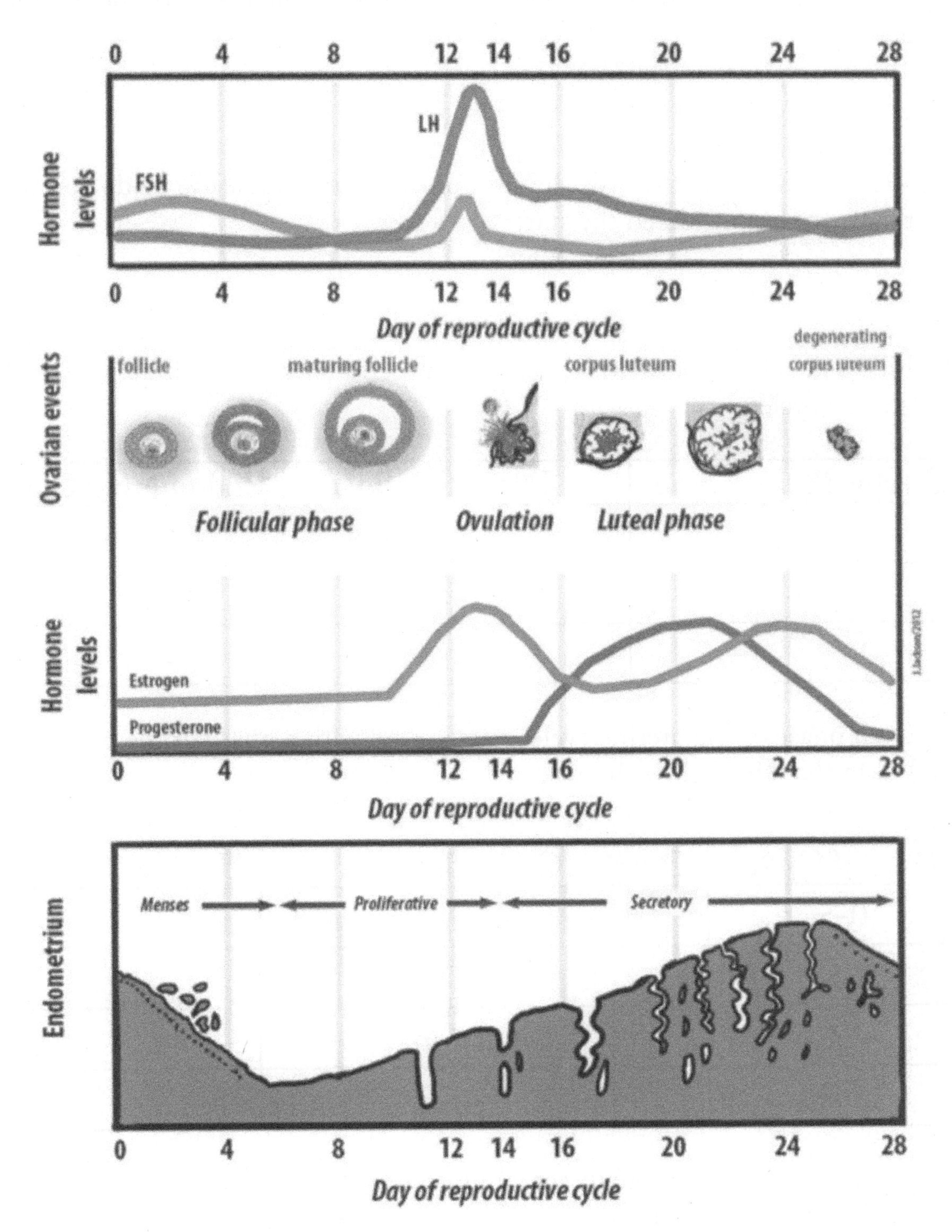

QUESTIONS:

17. What part of the uterus is shown in Model 4? Which layer of the uterine wall is this?

18. What uterine event signals the start of the uterine cycle? Why do you think this event is considered the start?

19. Fill in the chart below with information pertaining to the three phases of the uterine cycle:

Uterine Phase Name	Days of Cycle	Endometrial Changes

20. Each phase of the uterine cycle is associated with a phase of the ovarian cycle. Fill in the table below showing these associations.

Days of Cycle	Ovarian Cycle Phase / Event	Uterine Cycle Phase
0-5		
6-13		
14	Ovulation	
15-28		

21. Is the uterine cycle more directly influenced by gonadotropin or sex steroid levels? Explain your answer:

22. What is the relationship between the elevated levels of progesterone during the luteal phase and the thickness of the endometrium during days 16-23?

23. What is the function of progesterone?

24. Why is it important for the endometrium to increase in thickness and vascularity during the menstrual cycle?

25. When progesterone levels fall, what happens to the endometrium?

APPLICATION QUESTIONS:

26. If a woman has one ovary surgically removed, can she still become pregnant? Explain:

27. What are fertilization and implantation? What changes would occur in the uterine cycle if fertilization and implantation occur?

28. List three possible sources of reproductive failure (infertility) in females:

29. Review the information presented in Model 4 and explain possible mechanisms by which hormonal birth control methods work.

30. Do hormonal forms of birth control reduce one's chances of getting a STI/STD during sexual activity? Explain why or why not:

31. What is menarche? What causes menarche?

32. What is menopause? What causes menopause?